Ecology, Environment and Pollution

Ecology, Environment and Pollution

UMESH KUMAR

AMIGA PRESS. INC
Delhi -110051

AMIGA PRESS INC.
Registered Office:
19 DBS Building, Ansari Road, Darya Ganj, Delhi. 110002, India
Phone: +91-11-23283333 Email: amigapress@gmail.com

This edition is published by Amiga Press.Inc

ISBN : 978-81-929542-5-7

First Published 2018

Reprint 2026

Printed and bound in India by Replika Press Pvt. Ltd.

Preface

Ecology is the study of the "relationships among living organisms or between them and the physical environment." There are a number of approaches to a human ecology that have been applied since the early 1980s. These represent the increasing specialization in anthropology, not only by the subfields that were described earlier on, but also by different theoretical approaches. Some approaches, parallel those taken in the field of ecology, but with time lags of several years.

Environmental Ecology is a good text raises the problem of how environmental science should be taught. Many existing general texts focus on increasing environmental consciousness and teaching principles, often in simplified form, relevant to that goal. Improving the recognition of environmental issues among a wide segment of students is a legitimate need, even if the broad coverage of these texts can lead to a superficial science course. However, a legitimate need also exists to train students to work in environmental science. For this group, consciousness-raising should be secondary to rigorous development of scientific tools. Although Freedman leaves no doubt about this environmental advocacy, Environmental Ecology comes closer to scientific analysis in many chapters than do other texts. It assumes some background in ecology, biology, and chemistry, and case studies are rigorous.

Pollutants of streams, lakes and estuaries come from many sources. Excessive nutrients commonly originate in domestic sewage and run-off from agricultural fertilizer. Certainly the former is the major source of excessive nutrients in most streams and lakes. Toxic chemicals originate in industrial operations, acid waters from mine seepage or surface erosion, and washings of herbicides and insecticides.

Chemicals include such substances as pharmaceuticals, radionuclides, macronutrients such as phosphorus and nitrogen, trace gases and elements, particulates, and organic and inorganic compounds. They are used in, and directly or indirectly released from, processes and products that are essential for people's health, nutrition and well-being. However, chemicals and their breakdown products have hazardous properties which can pose a risk to the environment, ecosystem services and human health. Risk assessments, based on fundamental knowledge of transport, fate, exposure and effects, are essential for safe chemical use and release.

Environmental microbiology is the study of the composition and physiology of microbial communities in the environment. The environment in this case means the soil, water, air and sediments covering the planet and can also include the animals and plants that inhabit these areas. Environmental microbiology also includes the study of microorganisms that exist in artificial environments such as bioreactors. Microbial life is amazingly diverse and microorganisms literally cover the planet.

An average gram of soil contains approximately one billion (1,000,000,000) microbes representing probably several thousand species. Microorganisms have special impact on the whole biosphere. They are the backbone of ecosystems of the zones where light cannot approach. In such zones, chemosynthetic bacteria are present which provide energy and carbon to the other organisms there. Some microbes are decomposers which have ability to recycle the nutrients. Microbes have a special role in biogeochemical cycles. Microbes, especially bacteria, are of great importance because their symbiotic relationship (either positive or negative) have special effects on the ecosystem.

This book provides a systemic approach to understand this subject.

—Editor

Contents

1

Ecology and Climate Change

Ecology is the study of the "relationships among living organisms or between them and the physical environment."

Some characteristics of the scientific study of ecology:

- Macro-scale ecologists (global-scale ecology, '60's) have lost political ground to micro-scale ecologists.
- Landscape ecology recognizes human influence on non-human species, yet persist in distinguishing human from natural landscapes.
- Human ecology concentrates on systematic and evolutionary aspects, while social ecology emphasizes behavior. Both study human-environmenal relationships in distant past or present. Sociobiology is a paradigm within human social ecology
- Cultural ecology and cultural geography examine adaptive strategies. Both are cognizant of the role of culture in human adaptation, but not interested in long term change.
- Environmental History is the intellectual history of the environmental movement, including the political and economical implications of environmental interaction.
- Environmental Ethics explores value systems as they relate to human conduct.

None of these fields has truly integrative approach, and many lack an explicit historical component. What is needed is a multi-scalar temporal and spatial frame with an explicit focus on the role of human cognition in the human-environmental dialectic.

HISTORICAL ANALOGS

Global climate change is one of the most pressing event of current times. The anticipated changes demand investigations into patterns of human adaptation to climatic variability and change. However, the global climate change models used by physical scientists to predict climatic changes do not discriminate among biotic zones or anywhere near a human scale. Furthermore, many physical scientists assume that "novel circumstances" render any historical analogy to current anticipated global climatic change irrelevant. This attitude is due to:

(1) The lack of high quality long term (>100 yr.) instrumentally obtained data
(2) Local proxy data (such as tree ring) are only valid at the broadest temporal scales.
(3) Dismay of the comparative messiness of soft social science data
(4) Vested interest in favor of novel technologies and undervalue of traditional solutions

A regional approach could overcomes this. A region's air mass data, hydrology, soil, topography and species distribution can be used in regional models. Regionally documented ethnography, archeology, and documentary evidence evidences results of human activities and past choices which encompass the entire system. Multiple regional environmental changes can identify sensitive geographical locations. Interregional relationships may then be established and integrated with global data. This approach fosters creativity and the development of local and regional answers to global situations in which sensitive cultural issues play an important part.

INTRODUCTORY STATEMENT

We wish to trace in the barest outlines the movement in sociology which has led up to the ecological approach to sociological study. Sociology as a science was conceived when Vico hit upon the idea of producing general social principles and laws by studying and generalizing the data of history and literature. But neither Vico nor the important group of philosophers of history who succeeded him really created an inductive science of sociology. They were still under the spell of the metaphysical concept of natural law, which placed the determination of events in the past and in the distant bowels of the universe, or perhaps in the matrix-mind of divinity. It was Comte who played midwife to a theory of inductive sociology, by insisting upon the determination of events by the interaction and interrelation of the events themselves, instead of by factors or forces (natural law) outside of the events. Thus he stated the formula for all inductive science and named it positivism.

The inductive method had already begun to be used in the older sciences, although it had not been labeled and set in opposition to the older methods before Comte's keen penetration of the significance of intellectual history enabled him to draw the strong contrast between the positivistic and metaphysical approaches. The method of Vico, in the hands of the philosophers of history, had failed to make good its promise of providing social laws and principles. Historical data were not adequate to such generalization. Montesquieu had sought to remedy this defect by the introduction of geographic and climatic factors which he undertook with moderate success to generalize inductively. Buckle combined the methods of anthropogeography and the philosophy of history and elicited a great, if not a permanent, response.

But the death knell of a prioristic social thinking had already been sounded, and, interestingly enough, by the historians. They began to confine their attention to the collection, verification, and storing of facts, largely to the exclusion of their interpretation. All of the other social disciplines followed suit and began to place their main emphasis upon description and the historical method, which was for them a method of fact collection.

But they found it difficult as interpreters of contemporary life to avoid some sort of generalization. This was especially true of sociology. It was never possible after the downfall of the philosophy of history to make of sociology a mere collection of historical data. From its dilemma it found two means of partial escape. One of these was generalization by analogy-especially by biological analogy-in which morass it floundered for nearly a generation.The other escape was more logical and broke less definitely with the abstemious method of history. Sociology turned in the last third of the nineteenth century to ethnology for the materials it sought to use inductively.

But this recourse was not satisfactory, partly because the results were not much more dependable than those provided by the philosophy of history, and partly because there was an insistent demand that sociology should provide an explanation of life as it is. Already in the eighties the theological schools began to introduce courses in social problems, and this content was merged early in the nineties with sociology. Along with the content often went the clerical teachers themselves. Their neo-Calvinistic and English-ethical traditions led them rigorously to exclude materialistic factors from their accounts of causes, and even the economic factors they looked upon with suspicion and annoyance. For nearly two decades practically the only mention of physical and economic environment as factors permitted to enter into any serious discussion of social maladjustment came through the scarcely respectable writings of the Italian criminologists.

For the most part sociological theory was almost equally unfriendly to taking cognizance of material environmental factors. Recognition of the physical environment was relegated to geography, and the recognition of economic factors was either dismissed as socialistic or materialistic interpretation or gladly abandoned to economics, which itself shied at too close contact with this field.

With the exception of a few men like Sumner, the treatment of social processes was in terms of a neo-Hegelian or ethicohistorical ideology. The philosophers and theologues in our camp were numerous. The rise of the French school of neo-Hegelian psychosociologists in the nineties tended to confirm our own indigenous ideological trend in sociology. Sociology itself came frequently to be defined in terms of psychic interaction among persons, the whole stage on which the action took place being left out of account-at least in theory.

Of course, such an unrealistic sociology could not persist and it fell an easy victim, in its search to escape from a vacuum, to the instinct or biological deterministic interpretation, just as more than a generation before an earlier group of social thinkers, forbidden to generalize directly from insufficient data, had narcotized their imaginations with the biological analogy. But the seed of a realistic or environmentalist interpretation of human nature and social processes were never lost during these years of the Babylonish captivity of sociology by the shades of Calvin and Hegel. Thomas, at Chicago, in his wide range of reading and interests, brought his students in contact with McGee and the environmental anthropologists.

Sumner worked ahead with his subsistence theory of social origins. And above all, the historians, who in the first half of the nineteenth century, had led the way in the search for facts (which, however, carried the sociologists only as far as uncertain ethnological data), now began at the end of the century to return to nature. Turner and his school are as surely among the forerunners of the new sociology (I mean no offense to them) as are Comte and all the Germans Small lists in his most remarkable Origins of Sociology!

Sociology is today in the process of returning from a number of blind alleys, and it is still a bit dizzy. It is mostly back from its ideological and instinctivist excursions. But the spirit of Hegel still lives, especially through the social mirages of Durkheim and the cultural determinists.

In this last school the surviving influence of Hegel is strongest. A theory of culture which begins in culture and ends in culture, which knows no geography and can bear no environment except that inherited from the past or from some other place-and smothers even this under the term culture-is the latest refuge of the thwarted neo-Hegelian ideologists. The reason for this trend is, I think, easily found. Such a detached concept of culture is but another name for tradition. Thus the culture interpretation now so popular among sociologists is largely a new disguise for the old ideology and the worship of tradition.

Close aid to the "culture in a vacuum school" is the stratigraphic conception of science, which would lay down the subject matters and methodologies of the various sciences in layers, one above the other, somewhat after the manner of Comte's famous classification. Advocates of this viewpoint apparently would not allow any of the technique or content of one science to trickle through the cement and tile partitions to another. According to this school of thought a good sociologist should blush to recognize neurons, endocrines, selection (natural or otherwise), climate, or topography and contour. It is not strange, perhaps, that some of our more recently and less metaphysically trained sociologists should take a staff from the hands of the historians and venture to walk on profane ground.

Perhaps it is also not strange that the revolt should have taken an organized form most conspicuously among some of the doctors of just that department where the shade of Hegel has wandered most persistently and where the stratigraphic concept is perhaps strongest.

The new emphasis in sociology upon a realistic approach to its data and methodology is decidedly inductive and environmentalist in contrast to ideological. The term human ecology is largely accidental and is indicative of (1) the dynamic character of its environmental realism, and (2) of the avenue of its escape from the old vacuum sociology of the ideologists, by analogy with the more dynamic work of the plant ecologists.

The name may or may not last; it may be fortunate or unfortunate. But the realistic approach itself is to the study of the contemporaneous facts and factors which are actually shaping and conditioning human social behavior in an everchanging environment, material and psychic, physical and social. The papers that follow represent a few of the trend along which this new dynamic realism is working out an analysis of the interrelations of man and his environmental controls. If no papers in the conventional field of population analysis appear on the program of this division this year it is not because no such studies are being made. Perhaps no other phase of sociological research is now so well financed. But most of these studies are strikingly static.

They have not caught fully the spirit of the dynamic trend which lays emphasis upon the analysis of adjustment and changes of adjustment occurring along with changes in social equilibration. We shall not attempt to explain why the population studies have so seldom caught the dynamic impulse to study human and group adjustments which is so characteristic of the trend which calls itself human ecology. It may be because those who have planned or approved the studies have not caught the dynamic trend. Or it may be something else. But it is a satisfaction to observe in the new movement toward dynamic realism in sociology a promise of the fulfilment of Vico's conception and of Comte's formula for a science of society which should study actual facts of life inductively and from this study discover the social habits and organizations of men.

ECOLOGY WITHIN ARCHAEOLOGY

Within archaeology, interests in the environment date back to the 19th century. Eco-logical theory in archaeology tended to be linked to processes of culture change and evolution through the writings of Leslie White (1949) on unilinear evolution and Julian Steward (1955) on multilinear evolution. More recent interests in the 1960s and 1970s were in systems theory in behavior, and most recently, within the past decade or so, archaeologists have directed their interests toward regional climate change, historical ecology, and landscape ecology. It is also the case, as Karl Butzer (1990: 92) stated: that there has been an "..advantage of exposing archaeology to the intellectual cross-currents of anthropology. But it has also been disadvantageous, exposing archaeology to disciplinary fads and limiting effective contacts with other scientists."

ECOLOGY WITHIN BIOLOGICAL ANTHROPOLOGY

Biological anthropologists moved from 19th century and early 20th century typological approaches and race studies to the understanding of humans and their evolution via modern ideas about adaptation to the environment as a basis for understanding human variation. Ecological theory was tied also to evolutionary processes, but more in the realm of biobehavioral evolution and associated with the Darwinian concepts of "selection" and "adaptation to the environment" (Warren, 1951; Weiner, 1964). Ecological theory in biological anthropology became a fusion of evolutionary and ecological theory (Bates, 1953, 1960), along with ideas from environmental physiology (Dill et al., 1964), biogeography and human biogeography (Coon et al., 1950), demography (Spuhler, 1959), and human biology. Recently, socioecological theory has successfully been used to extend the synthesis between ecology and anthropology, focusing primarily on individuals rather than higher levels of organization.

Prominent in nearly all ecological theory in anthropology has been the concept of adaptation to the environment. Ecological studies in biological anthropology were stimulated in the 1960s by the work being done in ecosystems ecology by the scientists in the International Biological Program or IBP. At this time, there were several Human Adaptability Projects associated with the IBP that were influenced by systems science and efforts to modeling complex ecological systems.

TWO EARLY STUDIES

There were two very original anthropological studies that were done in the 1960s that were subject to a great deal of criticism in the anthropological literature. The first is the ecological study of the Tsembaga Maring of New

Guinea by the late Roy Rappaport. The work was done in the central highlands of New Guinea. The second is the energy-flow study of Andean Quechua Indians of Peru by Brooke Thomas. This work was done on the Peuvian altiplano at a base elevation of 4,000 meters above sea level. In the first case, Rappaport was a single investigator with an overwhelming task that he set out for himself. This is a long-established tradition in anthropology for one investigator to live with a people and, through participant-observation, to learn about the workings of the society or culture.

Rappaport not only took on the job of describing and understanding the inner workings of Tsembaga culture, he also attempted to understand this in the context of Tsembaga ecology. In the second case, Thomas's task was no less daunting, but his work was done within the framework of an integrated project. This was the Andean Biocultural Studies project, initiated by Paul Baker (Baker and Little, 1976) at the Pennsylvania State University in the U.S., with the primary objective of studying the patterns of adaptation of high-altitude natives to the hypoxia and cold of the Andean altiplano. When Thomas began his work, several years of data on social conditions, nutrition, human physiology, demography, and weather conditions had already been collected, and the area had been mapped.

Rappaport's Work

Rappaport's research was reported in a now famous book entitled Pigs for the Ancestors, which was published in 1968, and reprinted in 1984 with a 190-page Epilogue, in which he addressed his critics and reevaluated some of the research (Rappaport,1968, 1984). The work is brilliant, in that it addresses some of the fundamental issues underlying anthropological theory, including: social control, environmental causality for behavior, and the connection between individual behavior and cultural norms or prescribed social behavior.

In the work, Rappaport suggested first, that human population numbers, pig population numbers, the warfare cycle, agricultural productivity, patterns of exchange of goods, the distribution of land and people, and the maintenance of the ecosystem as a productive system were all tightly interrelated as a working system. Second, he suggested that the system was in a state of equilibrium maintained by feedback mechanisms. And third, and perhaps most controversial, that the regulating or controlling mechanism that kept the system going was the information provided in the form of ritual and a ritual cycle. Within this research, he took both a materialistic and a functionalist approach to social science, he identified human behavior as adaptive in the context of the social and ecological systems, and he identified human behaviors as subject to selection of favorable behaviors in the context of maintenance of the human/ecological system. Needless to say, and despite the fact that ecological anthropology was in vogue at that time, Rappaport's critics in sociocultural anthropology were severe in their verbal assaults. His work had attacked some of the fundamental icons of anthropology. He and other ecological anthropologists were accused of:

(1) reifying the ecosystem (to treat the abstraction of an ecosystem as if it had material existence);
(2) vulgar materialism (a belief that the materialistic approaches used in ecological anthropology were simplistic in their social context);
(3) a calorific obsession (placing too much emphasis on flows of energy through the system);
(4) excluding historical factors (too much emphasis on equilibrium and stability in diachronic state in the systems studied);
(5) setting up false boundaries (human cultures go beyond ecosystem boundaries);
(6) shifting levels of analysis (applying one level of interpretation to another); and
(7) dealing with an "impoverished" ecosystem approach (in contrast to "evolutionary ecology").

Scientists in the ecology community were debating some of these issues, but in many cases, anthropologists did not fully understand the bases for the debates, plus the human dimension added profound levels of complexity to these issues. In any case, the emotion behind these critical writings and the use of such intemperate terms such as "reification," "vulgar," "obsession," "false," and "impoverished" reflected the intense feelings about ecological and cultural materialistic approaches by a majority of anthropologists.

Thomas's Work

The work that Brooke Thomas conducted on energy flow research in a highland native community in Peru was begun in the late 1960s after Rappaport's and Vayda's ecological studies of New Guinea populations. The work was stimulated by Rappaport's and others' research and by the ecologist H.T. Odum's graphic shorthand language to represent the flows and controls of energy through ecosystems. At the time of the study, Quechua natives of the altiplano employed a mixed subsistence of cultivation of potatoes (and other tubers) and quinoa (chenopods) and herding of llamas, alpacas, and sheep.

By comparing food energy production (outputs) with labor expenditures (inputs), Thomas demonstrated that cultivation provided a 10: 1 return, while livestock herding provided only a 2: 1 energy return (Thomas, 1976). Animal products (meat, hides, wool) were highly prized at lower elevations; hence, trade of animal products for other foods (e.g., maize, sugar) increased the ratio to more than 7: 1.

Thomas's model, although representing averages and a simplified view of the energetics of production and expenditure in this community, nevertheless quantitatively demonstrated the utility of some of the principles of Quechua native subsistence through energy flow.

Thomas's work was the focus of an intense critique in a book called Energy and Effort that was edited by the distinguished human biologist Geoffrey A. Harrison (1982). The critique was penned by Philip Burnham, who began his comments by criticizing H.T. Odum's (1971) work on Environment, Power, and Society, identifying it as "reductionist" and "breathtakingly naive."

This book was somewhat naive, particularly in its chapters on human politics and religion, but many of the analytical approaches were very useful. Burnham (1982) continued his comments by outlining methodological problems that he saw as limiting understanding of human behavior by energy flow studies. One point has merit, where he stated: "...there is the problem of the multidisciplinary competencies required of a single researcher engaged in human ecological field study..." (Burnham, 1982: 233). Other arguments that he made were:

(1) that the costs are too high for the "pay-offs" of energetics (anthropologists have grown accustomed to very modest research budgets);
(2) adequate nutritional assessment is impossible from field studies (Michael Latham, an eminent nutritional scientist from Cornell University once told me that it was really the anthropologist who could address several key nutritional issues from extended field work);
(3) too many simplifying assumptions were made (this is a key to modeling, but only at the outset);
(4) it is impossible to account for all of the social issues (but, this is never even possible in sociocultural analyses);
(5) it is inappropriate to apply the functional/adaptational paradigm borrowed from biology (this reflects the hostile views toward the biological sciences that many social scientists feel). Adaptation as a concept was criticized heavily where he expressed his view on "...the inadequacy of the concept of adaptation as applied to social behavior!" In brief, Burnham typified the views of many sociocultural anthropologists (despite his interests in human ecology) where a materialistic, adaptational, quantitative approach that draws on basic biological principles somehow sidesteps the fundamental bases of human culture and society.

HISTORICAL ECOLOGY AND BIOGEOGRAPHY

A major outstanding question in historical ecology, and biogeography in general, is what role organismal characteristics play in determining composition of communities. Although it may seem obvious that organismal characteristics determine range limits of species, the issue of how those species came to be found in a particular region is open to question.

Interestingly, two completely independent intellectual traditions in Biogeography downplay the importance of organismal characteristics. The first is island biogeography (MacArthur and Wilson 1967), and its intellectual descendants (Hubbell, 1997,2001). This tradition considers raw species numbers, and considers individual species largely interchangeable. Although dispersal and extinction are both central to these theories, the difference in dispersal abilities between individual species are assumed to average out in the end.

The second great intellectual tradition is vicariance biogeography. Whereas dispersal is central to island biogeography, in vicariance biogeography the emphasis is on the signal of past fragmentation that can be gleaned from the species that do not disperse from their areas of endemism.

If species collected from the same areas show congruent phylogenies, this is considered evidence that they experienced the same series of vicariance events (geographic subdivision). Vicariance biogeography emphasizes huge events that affect entire biotas, regardless of individual organismal characteristics. In this sense, it is similar to island biogeography in that it seeks generality by downplaying organismal characteristics.

In contrast, a third view considers the possibility that organismal characteristics are important. Diamond's controversial assembly rules (1975) argued that the composition of many communities depends on organismal characteristics. In this vein, Cunningham and Collins (1998) have argued that phylogenetic history has the potential to reveal the influence of organismal characteristics in biogeographic histories.

If we are able to identify a set of species that have experienced a common biogeographic history, we can ask whether they share particular organismal characteristics. While this may sound at first like vicariance biogeography, it doesn't assume that there is a single overriding historical pattern (determined by vicariance) that defines an entire biota. Common patterns of dispersal, such as the trans-Arctic interchange itself, can generate patterns as well. In marine systems, with their porous barriers, it is unlikely in most cases that vicariance will affect more than a subset of the biota (with obvious exceptions such as the rise of the Isthmus of Panama). In many cases, we expect that subsequent dispersal between vicariated areas will be extremely common (Cunningham and Collins 1994).

With respect to the North Atlantic, this perspective can be framed in terms of two likely patterns. The first is glacially induced extinction in the NW Atlantic. If a species experiences local extinction without recolonization from elsewhere, the range of this species has been restricted. Vermeij(1989) pointed out that paleontology is the only discipline that can unambiguously identify cases of restriction. Cunningham and Collins (1998) argued that population genetic and phylogenetic data can accomplish the converse, and unambiguously identify sets of species that have long endemic histories in the NW Atlantic, and have therefore resisted glacial extinction.

Since repeated extinctions and recolonizations may appear in a low resolution fossil record as a coninuous presence, genetic data are important in this context. In high resolution mcrofossil records, these repetitive local extinctions and recolonizations can be directly observed. These stratigraphic records can then be compared to the conclusions of phylogeographic studies. The first observed pattern is one of long term residence on both coasts of the North Atlantic. Homarus gammarus in the NE Atlantic and Homarus americanus in the NW Atlantic are sister species, and have had independent evolutionary histories for roughly 1 million years or so (Tam and Kornfield 1998). This "reciprocal monophyly" is strong evidence that Homarus has been in residence in both the NE and NW Atlantic for at least 1 million years (Cunningham and Collins 1998). What is it about Homarus that has allowed it to resist extinction in the NW Atlantic through several glacial maxima? The second is a population genetic pattern shown by many species strongly suggesting post-glacial colonization of the NW Atlantic region from the NE Atlantic (e.g. the seastar Asterias rubens, and the snails Littorina obtusata, and Nucella lapillus (Wares and Cunningham in revision)). What characteristics allowed these species to invade across the Atlantic? Since the latter two species have no pelagic dispersal, dispersal ability is not the only important factor in colonization ability. Other factors inlude the ability to withstand the wide temperature fluctuation in the NW Atlantic, and the tendency to occupy dispersal vectors such as floating algae (Ingólfsson, 1995).

ECOLOGICAL RESEARCH IN ANTHROPOLOGY

We believe that the answer to the question posed in the title of an earlier talk on whether there is a future for ecological studies in anthropology is "yes," but probably still at the margins of anthropology, not in the mainstream. Carole Crumley (1998: ix-x), an archaeologist with interests in historical and landscape ecology, observed that the anthropological subdisciplines most closely allied with the sciences-archaeology, biological anthropology, and human ecology within sociocultural anthropology-have been marginalized in anthropology for much of the latter half of the 20th century.

These are also the subdisciplines most closely allied with environmental sciences, and, hence, are more receptive to ecological approaches. There are several areas of exploration within anthropology that would profit from ecological applications and multidisciplinary collaboration. These include studies of: landscape ecology within sociocultural anthropology, historical ecology within archaeology, urban ecology and managed ecosystems, within anthropology, broadly, and biodiversity and global studies, again, within anthropology, broadly.

Landscape Ecology within Sociocultural Anthropology

There is considerable interest in land use in the Third World by sociocultural anthropologists and other social scientists, particularly in the context of human population growth and increasing pressure on soil and land resources. There are disagreements over whether pastoralists and cultivators contribute to desertification, and how cultivation of the land can be best managed in semiarid or wet tropical lands. I recall an important observation that George Innis made in modeling slash-and-burn agriculture in the tropical rain forest more than 25 years ago. Innis (1973) reported that repeated use of tropical swidden plots, even with up to 40 years of fallow between each period of cultivation, would lead to depletion of several essential soil nutrients (potassium and soil organic matter), and that some of these nutrients required more than a hundred years for full recovery.

What struck me was that conventional anthropological wisdom dictated that the indigenous pattern of eight years fallow was sufficient to maintain sustainability indefinately. The value of this pattern of use was embedded in the anthropological literature, yet it was clearly false. In this case, so much could have been gained by collaboration.

Historical and Landscape Ecology within Archaeology

Historical ecology, as a new paradigm, has been embraced by archaeologists with materialist and environmental interests. Since the landscape and landscape transformation are central concerns, the ecological conditions of human history must be understood within the context of the cultural conditions for a comprehensive interpretation. Some consider it "..the most important current intellectual advance in the study of human and environmental relationships" (Balée, 1998: 2).

I would moderate this statement a bit, but I agree that this approach does hold great promise for integrating archaeology, ecology, and history in interesting ways. Anthropogenic landscape alteration in prehistoric times is well documented, especially in the New World before Columbus. Terrace, canal, and road construction in the Andes produced dramatic transformation of the land-scape, and the lowland Amazon forest was modified by slash-and-burn cultivation and by raised fields. Raised-field cultivation was a widespread agricultural technique and used throughout the lowlands of Central and South America. A recent application of this kind of landscape archaeology is in the exploration of raised cultivation fields in the seasonally flooded areas of the Bolivian Llanos de Mojos (Erickson, 1995). Techniques used in the Llanos de Mojos work, in addition to archaeological methods, included agro-climatic modeling, ethnobotany, remote sensing, and experimental construction. Few people use these raised fields today in South America despite their effectiveness and the apparent sustainability of this indigenous pattern of cultivation.

Urban Ecology and Managed Ecosystems

One of the dominant trends in human populations over the past century has been the movement of people from rural to urban settings (Bogin, 1988). Rural-to-urban migration takes place largely because individuals perceive that cities are centers of economic opportunity and excitement. The process, whether within or between national boundaries, has contributed to remarkable urban growth and widespread conditions of congested living that are unprecedented in human history. The beginning of the 21st century was marked as the time in which more than 50 percent of the earth's population were living in cities. Migration from the countryside to the city dates back to the rise of cities in antiquity (McNeill, 1978), and has been one of the most common types of migration since that time. Indeed, until the last century, urban mortality rates were so high that most cities could not even maintain their sizes, much less grow, without substantial numbers of immigrants (McNeill, 1979).

The linkages between rural-to-urban migration, demography, epidemiology, and urban ecology are crucial ones if we are to understand this highly-modified urban ecosystem. The urban LTERs in Phoenix and Baltimore, as well as other urban studies around the world, can serve as test cases. If the topical mix is demography, epidemiology, and urban ecology, the disciplinary mix should certainly include the social, biomedical, and ecological sciences.

Biodiversity and Global Studies within Anthropology

Within the past two decades or more, ecologists have become increasingly aware of losses in numbers and kinds of organisms around the globe and in alterations in the biosphere. At the same time, climatologists and other scientists tracing global trends have identified disturbing patterns associated with anthropogenic effects on the planet (e.g., increased atmospheric CO^2 and pollution, progressive soil loss). Since anthropologists, particularly biological anthropologists, are more closely attuned to environmental effects on humans, they have been slow to become involved in exploration of biodiversity losses. Another limiting factor to the participation of anthropologists in assessing changes in the biosphere is the problem of spatial scale; that is, such large-scale, global problems are usually outside the scope of anthropological investigation.

One international program in which anthropologists and other social scientists might participate as members of multidisciplinary teams is DIVERSITAS, an international program of biodiversity research, consisting of 11 research components (Diversitas, 1996). The "Human Dimensions of Biodiversity" component of DIVERSITAS incorporates human-oriented disciplines within the general themes of the other components and is designed to contribute to an integrated approach to the study of losses in biodiversity. Most recently, the U.S. National Committee for the International Union of Biological Sciences (IUBS) prepared a document that defines how some social scientists and human biologists might contribute to DIVERSITASefforts (Little et al., 2001). The U.S. National Committee for the International Union of Anthropological and Ethnological Sciences (IUAES) will also contribute to the definition of a U.S. Program in this area.

There are several essential areas in which anthropologists might contribute to these very important studies. First, there is the area of human impacts on biodiversity, which is certainly the central feature of biodiversity losses. The impacts of human populations in transforming the landscape, competing for habitats, contributing to pollution, and outright predation, are profound, indeed. Among many kinds of studies, it is here that archaeologists and ethnohistorians can document some of the long-term changes in biodiversity and how rates of loss have varied through time. Second, how has human biodiversity changed in the context of other species' losses? This is a vast area of exploration, but one that might draw human biologists into DIVERSITAS research. In a third area of exploration: how do human perceptions of biodiversity influence our abilities to respond to losses or to take action? How does culture play a role in these responses?

DIVISIVE ISSUES IN ANTHROPOLOGY

In 1968, Vayda and Rappaport stated that: "...a unified science of ecology has definite contributions to make towards the realization of anthropological goals and does not entail any appreciable sacrifice of traditional anthropological interests". There are a number of reasons why this objective has been achieved only in small measure that relate to some fundamental traditions in anthropology.

Following the Second World War, as with many other sciences, there was an increasing specialization by subfield accompanied by a tension between sociocultural anthropology and biological anthropology. Part of the basis for this was that in the 19th century and early 20th century, physical anthropology was preoccupied with race studies and there were clear racist elements in many of these studies.

This tension between "social" and "biological" intensified during the second half of the 20th century, when social scientists were concerned about "biological" and "genetic determinism" and other paradigms that placed heavy emphasis on human biological processes taking precedence over behavioral and social processes. Asuspicion by social scientists of all biologically based paradigms arose on the one hand and was paralleled on the other by a need to defend fundamental social processes and theory on the other. This "biophobia" is by no means universal in the social side of anthropology, but it does play an important role in the acceptance of certain ideas. Another tension that divides anthropology is the difference in approach between the scientists and the social humanist/historian (materialist interpretation vs. a symbolic/cultural interpretation). Beliefs that human social behavior is so complex that is can never be fully understood by conventional scientific approaches are quite common among anthropologists, and, in fact, limit attempts to systematically study human behavior.

Some time ago, postmodernism entered anthropology via literary theory with challenges against the fundamental value of systematically-gathered information, and even objective reality. Within the social humanistic side of anthropology, there is a strong interest in "praxis" or practice or applied anthropology in reducing the effects of poverty in the Third World as well as in Western nations. Associated with these applications of anthropological knowledge is a kind of "anthropocentrism" (Rappaport, 1984: 387), which places the environmentalists and scientists who are concerned with the conservation of nature at odds with the social scientists, who see the world filled with poverty that has arisen in part because of differential knowledge and what is known as unequal power relationships. In this context, the social scientists are inclined to follow the idea of "putting people first," also the title of a successful collection of papers of development in the Third World (Cernea, 1991). This struggle between human needs and the need to maintain viable ecosystems is an exquisite conflict with an uncertain outcome (Newmark and Hough, 2000).

It is an area of investigation where collaboration between natural and social scientists is urgent! Another issue is an extraordinarily complex one: that of Garrett Hardin's "Tragedy of the Commons" (Hardin 1968). Agreat deal has been written about this issue, both within and outside of anthropology.

The fundamental objection that the anthropologists have against Hardin's basic premise is that it violates the concept of human agency; that is, the ability of humans to manage their own environments, and through cooperation, to avoid the tragedy of environmental exploitation that Hardin described. This belief in human agency is also linked to the unwillingness of many anthropologists to even entertain the idea that human behavior can be influenced by the circumstances of their environment.

An issue of the Social Science Research Council Items & Issues newsletter (Wissoker, 2000) was devoted to an article and commentaries on "advancing interdisciplinary research." In the lead-in to the collection, the editor noted: "...indeed, the idea of interdisciplinarity was practically born here..." What is significant about the contributions is that all of the commentary is by social scientists, and there is absolutely no mention of health, disease, the environment, ecology, or any of the natural sciences in their schemes of interdisciplinarity! These social science approaches to human ecology, are what in anthropology was called "cultural ecology." This approach, with an ecological emphasis on sociocultural process within the context of current anthropological theory was reviewed 25 years ago by Orlove (1980) and more recently by P. Little (1999).

Despite the bleak picture I have painted of social science and of anthropology in the context of science and ecology, some good work has already been done, and I am hopeful that new programs of collaborative and multidisciplinary research can be initiated between ecologists and anthropologists. However, it should be emphasized again that it is probably impossible for a single anthropologist or a single ecologist to conduct a study of human ecology and reach meaningful conclusions. The tasks are too vast for single scientists working alone and the solution is, of course, to establish multidisciplinary projects. Some examples of earlier and ongoing research and prospective ecological studies can be discussed.

INTEGRATED STUDIES OF SINGLE POPULATIONS

Beginning in the early 1960s, at the time that the International Biological Programme (IBP) was being organized, a number of single-population integrated projects were begun. Most of the projects were identified as a part of the Human Adaptability component of the IBP and were initiated with the concern that these populations were endangered, and their extinction would mean the loss to science of populations that most closely resemble human populations during the greater part of our evolutionary past. Later projects in the same pattern of investigation were started in the 1970s, some under the Unesco Man and the Biosphere (MaB) program. Human biologists or biological anthropologists organized most of these projects, but some were integrated with social scientists playing key roles.

There were several themes that these projects represented, including: adaptation to the environment, in its broadest sense; microevolution; cultural and biobehavioral evolution; health, epidemiology, and culture change; and ecology and systems science (Little et al., 1997). Some of these major integrated and multidisciplinary projects are listed here. Adaptation to (1) arid environmental conditions and limited resources in Kalahari hunter-gatherers; (2) high-altitude hypoxia and cold in Andean Quechua; and (3) Arctic cold in circum-polar Siberian, Inuit, and Algonkian populations, were studies conducted within a framework of populations living under the stress of extreme environmental conditions.

Microevolutionary studies were conducted of the genetics of the Amazonian Yanomama, Makiretare, Cayapo, and Xavante, the Andean Aymara, the Central American Garifuna, Solomon Islanders, and several populations of central African Pygmies.

Language, genes, demography, culture, and phenotype, were used to explore ongoing evolutionary processes in these populations, and to reconstruct processes in the past. Attempts to reconstruct cultural and biobehavioral evolution of the paleolithic were made in the Kalahari and Pygmy studies. Health, epidemiology, and culture change were central issues in the Circumpolar, Tokelau Island Migrant, Samoan Migrant, and several other projects. Here the effects of modernization on native populations were a primary objective. Finally, although an interest in the influence of the environment on a population's behavior and biology was a common theme in all of these integrated projects, only a few had real interests in ecology and systems approaches (Little et al., 1991). As already noted, Thomas (1976) carried out energy flow modeling on Andean Quechua farmer-herders. Later modeling focused on attempts to explain why some nutrient sources are important and why others are not. Gage (1986) applied optimal foraging theory to the slash and burn agriculture of the Samoans and identified that the net rate of energy production (NREP) was indeed a central guiding principle when Samoans considered production of their three primary crops: breadfruit, banana, and taro. Hett and O'Neill (1974) developed a carbon flow model for Aleuts that demonstrated a heavy dependence on marine organisms and the need to incorporate terrestrial and marine ecosystems in the analysis of Aleut food webs.

Finally, systems ecology figured prominently in the Kenya Turkana research because that was a central approach taken, but also because ecologists and anthropologists worked closely together. The synthesis of this work concluded that Turkana pastoralists were capable of surviving and flourishing in a dry and highly variable ecosystem by complex livestock management, mobility, opportunistic exploitation of resources, and adaptive social patterns of sharing-while at the same time, avoiding degradation of the ecosystem (Leslie et al., 1999)

ECOLOGY AND THE POLITICS OF KNOWLEDGE IN MODERN SCIENCE

The paradigm of modern science has evolved in the last few centuries in an environment where all economic activities were aimed at maximising the productivity of man-made processes in individual sectors of the economy. This led to the development of modern technologies with highly negative externalities which remained invisible within the conceptual framework of modern science and economics. This shortcoming emanates from three basic fallacies of modern scientific knowledge:

1. It identifies development merely with sectoral growth, ignoring the underdevelopment introduced in related sectors through negative externalities and the related undermining of the productivity of the ecosystem.
2. It identifies economic value merely with exchange value of marketable resources. Ignoring use values of more vital resources and ecological processes.
3. It identifies utilization merely with extraction. Ignoring the productive and economic functions of conserved resources.

Development planning based on these false identifications tends to create severe ecological problems because of its inability to recognise ecosystem linkages and the ecological processes operative in the natural world. The ecological relationships between the sectors of natural resources contribute to essential ecological processes which are frequently found to be vital for human survival. Thus, the stability of ecological processes is not merely a matter of aesthetics. An incomplete understanding of the material and economic values of ecological processes leads to the destruction of the material conditions for economic development and eventually survival. Since the availability of essential and vital resources for survival is dependent on the maintenance of essential ecological processes, economic activities which generate sectoral growth in the shortterm by destroying the essential ecological processes cannot lead to development in the long run. On the contrary, by decreasing the productivity and availability of vital resources, they initiate the process of underdevelopment.

When the natural world is viewed ecologically as a system of interrelated resources which maintain the material basis for human sustenance, economic values can no longer be perceived merely as exchange values in the market. Economic values in the ecological perspective are not always equivalent to their exchange value in the market, evaluated without any significance to their use value.

As a corollary, natural resources can have economic utility that cannot be quantified through the exchange value in the market. Such economic utility includes the maintenance of essential ecological processes that support human survival and, thus, all economic activities. The economic utilisation of resources through extraction may, under certain conditions, undermine and destroy vital ecological processes leading to heavy but hidden diseconomies. The nature of these diseconomies can be understood only through the understanding of ecological processes operating in nature.

The economics of sustenance and basic needs satisfaction is, therefore, linked with ecological perceptions of nature. The economics of sectoral growth on the other hand is related to reductionist science and resource wasteful technologies which are productive in the narrow context of sectoral and labour inputs, but may be counter-productive in the context of the overall economic base of natural resources.

The representative of thousands of such cases seen everywhere. They reveal a certain pattern of contemporary economic development which can be identified thus:

1. Development has been equated only with the growth of manufacture in individual industrial sectors and with the increase in productivity of only man-made processes.
2. This sectoral growth of man-made processes has also led to ecological destruction of the natural resource base, affecting negatively other sectors of the economic system. This leads to the decay of systems productivity of all productive processes, man-made and natural.

As a result of this limitation of contemporary economics, economic development has, consequently, been taken to be synonymous with growth. The higher the rate of sectoral growth, the higher is the index of economic development. Possible ecological destruction caused by the resource intensity of sectoral growth that is guided purely by non-ecological economic considerations, has never been introduced in the processes of planning for economic development. The benefit-cost analysis of development projects has thus externalized those ecological changes and is incomplete in three important ways:

1. It deals with benefits and costs as profits and losses in financial ferms.
2. It deals with benefits and costs only in the narrow sectoral perspective and ignores costs generated by inter-sectoral linkages.
3. It deals with benefits that are largely available to more visible and economically powerful groups and ignores costs that are borne by the less visible and economically weaker groups. These costs and the

associated underdevelopment are thus made invisible in modern economic analysis.

The utilisation and management of natural resources in India has so far been guided by the narrow and sectoral concept of productivity and restricted benefit-cost analysis. This narrow concept of productivity and benefit-cost analysis has blocked the conceptualization of the criteria of rationality of technology choice which maximizes needs satisfaction while minimising resource use, thus maximising systems productivity.

Natural Forests in the Catchments of Rivers

For example, the clear felling of natural forests in the catchments of rivers, and planting of industrial species of trees has been justified on the grounds of increasing productivity of forests. This concept of productivity is, however, only related to productivity of industrial timber, while forests produce other forms of biomass, like fodder and green mulch, or maintain productivity of soil and water resources. The direct impact of the clear felling of catchment forests on agricultural production through its destructive impact on soil and destabilisation of the hydrological balance is not taken into account in the calculations of the benefits and costs associated with forests. Regular floods and droughts, which are the consequences of irrational land and water management, are branded as natural disasters for which the whole nation pays heavily. Consequently, the poor and marginal groups which depend on agriculture for their livelihood face increasing impoverishment and poverty. This thrusting of negative externalities on the poor and marginal groups directly leads to the polarisation of society into two groups. One group gains from the process of narrow sectoral growth, while the poor and marginalised majority suffer because of the ecological destruction of natural resources on which they depend for survival.

The dialectical contradiction between the role of natural resources in production processes to generate growth and profits and their role in natural processes to generate stability is made visible by movements based on the politics of ecology. These movements reveal that the perception, knowledge and value of natural resources vary for different interest groups in society. The politics of ecology is thus intimately linked with the politics of knowledge.

For subsistence farmers and forest dwellers a forest has the basic economic function of soil and water conservation, energy and food supplies, etc. For industries the same forest has only the function of being a mine of raw materials. These conflicting uses of natural resources, based on their diverse functions, are dialectically related to conflicting perceptions and knowledge about natural resources. The knowledge of forestry developed by forest dwelling communities therefore evolves in response to the economic functions valued by them. In contrast, the knowledge of forestry developed by forest bureaucracies, which respond largely to industrial requirements, will be predominantly guided by the economic value of maximising raw material production.

The way nature is perceived is therefore related to the pattern of utilisation of resources. Modern scientific disciplines which provide the currently dominant perspectives of nature have generally been viewedes 'objective', 'neutral' and 'universally valid'. These disciplines are, however, particular responses to particular economic interests. This economic determination influences the content and structure of knowledge about natural resources which, in turn, reinforces particular forms of resource utilisation The economic and political values of resource use are thus built into the structure of natural science knowledge.

SIGNIFICANCE OF THE DISTRIBUTION OF AMAZONIAN DARK EARTHS

The total geographical extent remains unknown. Various authors summarize new information about the geographic distribution of ADE in Amazônia from systematic archaeological and pedological surveys. Sombroek et al. (2003) estimate that ADE covers 0.1-0.3% or 6,000-18,000 km^2 of the total Amazon Basin (6 million km2). ADE is reported for most of the major rivers of the Central and Lower Amazon Region. Most are associated with bluffs overlooking várzea near larger active or abandoned river channels and a few have been found on terra firme away from main rivers. Dispersed ADE are reported from certain riverine locations in the Upper Amazon of Peru, Ecuador, and Colombia.

Kern et al. (2003) cite recent surveys documenting ADE "every 5 km along the igarapés, and an over-all spatial coverage of one per 2 km^2". In the Upper Xingú region, Heckenberger (1998) estimates large ring-plaza villages, many of which have ADE, are spaced several kilometers apart. In the Upper Madeira river, ADE is distributed 1 per 2 km^2 (Miller, 1992: 220). High densities of ADE are reported for the Central Amazon Basin. Although the best-known ADE are large (ranging from 500 ha for the Santarém site, 200 ha for the Belterra site, 80 ha for the Manacapuru site; and 90 ha for the Altamira site; Smith, 1980, 1999; Denevan, 2001 and authors of this volume), 80 % of the known ADE are less than 2 ha (Kern et al., 2003). The densest distribution and largest ADE are associated with archaeological settlement sites along the middle and lower courses of the major rivers or "on the margins or confluence of streams and rivers or near falls" (Kern et al., 2003). As discussed above, in the past scholars have highlighted the distinctions between the potential resources of the larger floodplains (várzea) and interfluvial or upland regions (terra firme) which was reflected in the archaeological record as larger, more permanent settlements and associated with more complex societies established along major rivers and less permanent settlements and simpler societies associated with interfluvial regions. According to these scholars, the large, dense populations of the várzea were sustained by fish and other aquatic resources and farming the rich, annually renewed, floodplain soils of white water (sediment-rich) rivers. Thus, it is not surprising that ADE would be associated with large rivers.

In the 1980s, scholars began to challenge the assumption that the interfluvial uplands (terra firme) were homogeneous and resource poor. Thus, finding ADE in terra firme is not surprising. Although inland, most terra firme ADE are on bluffs above smaller upper tributaries and streams. These ADE sites are generally smaller and more dispersed than those associated with larger rivers (Smith, 1980). The large inland ADE the Belterra Plateau between the Tapajos and Curua rivers and along the Arapiuns river reported in this volume and by Smith (1999) are prominent exceptions. Nimuendajú (1952: 11) reported terra firme ADE associated with deep and wide artificial wells.

In an important revision of the floodplain vs. interfluvial (várzea vs. terra firme) hypothesis about Amazonian cultural development, Denevan pointed out that most prehistoric, historic, and present large settlements are located on the terra firme bluffs overlooking an active (or what was at one time) channel of the river rather than on the floodplains of the Amazonian drainage. The Bluff Model highlights the advantages of this location: direct access to floodplain and interfluvial resources, canoe transportation, defense, and dry locations for year-round settlement.

Denevan also points out that most pre-Columbian sites on bluffs are ADE, often surrounded by terra mulata. The large ring plaza villages of the Upper Xingú River are examples of bluff ADE over-looking floodplains in the smaller upper drainages (Heckenberger et al., 1999). As Kern et al. (2003) state, "...ADE are present in practically all types of eco-regions and landscapes," but based on their maps and those of others, the largest and most numerous are found adjacent to lower and central courses of larger rivers (supporting the earlier interpretations of Carneiro, Lathrap, Denevan, Roosevelt, and others regarding the ecological advantages of access to rivers over inland locations for settlement and cultural development). Returning to distribution of ADE, here we would like to stress two points 1) ADE have not been found everywhere within the Amazon Basin and 2) long-term permanent human settlement does not necessarily result in ADE. ADE are less common the Upper Amazon of Peru, Ecuador and Bolivia, Rio Negro drainage, Orinoco drainage, the Llanos de Mojos of the Bolivian Amazon, and the northern part of the Amazonian drainage basin. Why did ADE form in some areas and not in others? Is the absence of ADE a product of differences in classifications of soils, lack of archaeological and soil survey in these regions, or burial, erosion, leaching, and destruction of ADE sites? Denevan (2001: 114, footnote 4) suggests that lack of reported ADE in the Western Amazon may be due to "lack of awareness".

Only a small sample of the Amazon Basin has been systematically surveyed for archaeological sites; and future research will probably identify new ADE in these regions. On the other hand, many scholars are now familiar with the concept of ADE and actively looking for sites; thus, the geographical distribution may be real and significant. Did these areas originally have ADE that disappeared because of overexploitation, poor management, and/or natural processes of decomposition, erosion and leaching? Although unlikely, these possibilities should be addressed. An examination of the patterning of spatial and temporal distribution and cultural traditions where ADE is found provide some insights.

Does settlement size and duration play a role in whether or not ADE is formed? Most scholars seem to agree these are important factors. Were the cultures of regions without ADE less evolved and lacking in socio-political complexity compared to those with ADE? A brief examination of two areas that share a similar site plan but not ADE is instructive. The ditched enclosures or ring-ditch sites are elegant patterns of ditches and embankments covering several hectares to several square kilometers located on forest islands and river bluffs in the NE Bolivian Amazon. The Bolivian sites are probably part of larger related cultural phenomena of ring plaza settlements associated with complex societies reported for the Upper Xingú, Guaporé/Itenez, and Madeira river drainages. ADE is associated with many of these sites on the Upper Xingú, central Madeira, and Guaporé/Itenez rivers while those of northeastern Bolivia do not have ADE.7 The present day border between Bolivia and Brazil (the Guaporé/Itenez River) may mark a cultural boundary. This also roughly marks the present day transition between high canopy tropical forest and savanna. The Bolivian sites have not been adequately dated but appear to be late prehistoric. The Brazilian sites have a long chronology; and thus, ADE may have formed through longer continuous occupation.

Could the presence and absences of ADE be associated with differences in terms of natural resources, environments, soil types, agricultural practices, or settlement type? These geographical distributions suggest that the differences may be cultural rather than environmental. Basic cultural explanations may account for ADE. Organic matter placed directly in fields and "used up" under cultivation vs. organic matter accumulated to form ADE for later use as farmland are significant farmer decisions that could determine the presence or absence of ADE. In some cases, Amazonian riverine communities dispose of garbage by tossing it into the river. Stocks (1983) reports that the Cocamilla of Peru discard garbage into local lakes that is said to increase the populations of the fish they consume. In contrast, the Tukanoans of the Colombian Amazon (Chernela, 1982) and the Ka'apor of eastern Brazil never dispose of garbage in the rivers (Balée, 1994). A simple decision about garbage may determine whether settlements produce or become ADE or not. In cultures where garbage was tossed into rivers, streams, and lakes, no terra preta-type ADE would form; in others where garbage was deposited and accumulated on or near the settlement, terra preta-type ADE could form.

DEFINING AMAZONIAN DARK EARTHS

The authors in this volume are in agreement that the basic characteristics of ADE are their dark color, richness in charcoal-derived carbon, high fertility, and human origin. Beyond this, there is little agreement. The diversity of approaches for characterizing and identifying ADE are highlighted in this volume. The authors concur that ADE should not be defined too narrowly because of the rich variation within and between ADE (Kämpf et al., 2003). The lack of a single definition or clearly defined suite of characteristics for identifying ADE is obviously frustrating for some of the participants in this volume. At this point in the study, flexible definitions that recognize the variation are healthy.

How is ADE soil identified? The principle criterion is dark color. Various issues can be raised about color. How dark do soils have to be in order to be identified as ADE? Is a simple threshold on a Munsell color chart reading sufficient? Could soils be of radically different colors but have the same composition and formation process? What "natural" soil benchmark is used for comparison? At what point does archaeological soil (soil of archaeological settlements, monuments, middens, and earthworks) become ADE? Are ADE without the presence of archaeological artifacts anthropogenic? The authors of this volume grapple with these questions and provide some answers. In an attempt to address the issue of continuous distribution of soil color from brown or gray to black, scholars working in the central Amazon Region (Sombroek, 1966) introduced the term terra mulata ("brown soils") for the large transitional zone around terra preta (the classic ADE).

The diverse meanings of ADE and potential problems of communicating between scholars of different disciplines were clearly represented in the 2002 TPA Workshop and in this volume (Kämpf et al., 2003). By relying on a vague definition of ADE, we potentially open ourselves to charges of over-or underestimating the geographic extent and importance of ADE. On the other hand, a overly narrow definition might ignore the rich variation of ADE reported in this volume. Kämpf et al. (2003) consider the dynamic, historical and variable nature of ADE as a subset of general anthrosols (soils produced through human activities) for their Archaeo-pedological Classification. Their new classification is an attempt to combine insights from various disciplines to address the variability and continuous variation of ADE.

Archaeologists, who identify and map ADE, tend to rely on the discipline's soil classifications and interpretations that rarely agree with those defined by modern soil science. Archaeologists are experts at recognition and interpretation of a wide range of anthropogenic soils associated with human activities on sites and landscapes. Soil scientists focus on the horizons in the profile and consider anthropogenic features as noise, perturbation and disturbance. In contrast, archaeologists define the visually and/or texturally obvious anthropogenic features in the profile and treat the horizon formation as noise, perturbation, and disturbance. To the archaeologist, "natural" soils are only interesting in terms of defining the boundaries of the anthropogenic soils (i.e. site boundaries; sterile boundary under site, and so forth). Archaeologists are most likely to focus on the internal variation of ADE: faint patterns, changes of texture, color, context, fill, and features to extract function and meaning. Without the internal heterogeneity within a site, we could not do archaeology. Soil scientists are less concerned with these soil nuances and attempt to characterize features representative of larger spatial areas. The approach proposed by Kämpf et al. (2003), and to a certain degree, traditional archaeology dedicated to building chronologies, focuses on profile descriptions of small, often deep, excavations through ADE (or sites) which emphases vertical continuity and disjuncture. Since the 1970s, archaeologists have used large areal excavations often combined with sampling to recover spatial patterning of human activities and lifeways throughout the site, emphasizing a horizontal perspective. Kämpf et al. (2003) discuss the contrasting approaches used by archaeologist and soil scientists in regard to ADE and highlight the benefits of combining both approaches. ADE research can benefit from both the general chronological approach and spatial morphology of settlement and agriculture approach.

The practice of traditional archaeology is framed within the site concept. A site is a basic discrete unit of analysis defined by concentrations of artifacts indicating settlement or other activity assumed to reflect human behavior. I argue that adherence to the "site concept" limits our understanding of historical ecology in the Amazon Basin. The concept of landscape within Historical Ecology and archaeology of landscape are powerful alternatives to site-based approaches and are what links archaeology to historical ecology. Rather than focus on arbitrarily defined sites, landscape approaches try to understand human activities that occur between traditional sites and across larger areas at multiple scales. In this perspective, human activity is viewed as continuous over the landscape rather than spatially contained within traditional sites. Despite new innovative methods for archaeological survey, recovery of human residues, and "non-site" landscape approaches for defining between-site human activities (e.g. Stahl, 1995), Amazonian archaeologists are drawn to "sites," usually pre-Columbian settlements, defined by conspicuous surface concentrations of pottery, lithics, and charcoal (the most commonly preserved archaeological materials).

By extension, ADE research has adopted the site concept. Are ADE discrete spatial units of analysis as presented in this volume? How can a typical black earth site be measured if it has no clearly defined boundaries or edges? Most archaeologists and historical ecologists now recognize that the earth's surface is covered with continuous distribution of artifacts and evidence of human transformation of the landscape, making it difficult, if not impossible to clearly define boundaries. For example in the agroforestry literature on the Amazon Basin, every landscape has been transformed to some degree by thousands of years of human activities (burning, selection for economic species, weeding, and artificial disturbances).

The contributors of this volume often contrast ADE with the surrounding forest soils based on the assumption that the ADE are anthropogenic and the forest is "natural". What if the benchmark against which ADE is identified and defined is also anthropogenic? If the entire Amazon Basin is to some degree anthropogenic as some historical ecologists argue, the possibility of finding a totally pristine natural soil in Amazônia after thousands of years of human disturbance for comparison as a benchmark is unlikely. The concept of domestication of landscape (Clements et al., 2003; discussed below) may provide an alternative to the site concept. Most ADE discussed in this volume are entire sites or a subset of traditionally defined sites (most covering hectares).

The boundaries of an archaeological site and its ADE do not always correspond such as in the case of the Açutuba site on the Rio Negro where surface artifact scatters much larger than the ADE (Heckenberger et al., 1999) or the Araracuara sites on the Caquetá river where ADE in the form of terra mulata extends far beyond the distribution of artifacts (Mora et al., 1991). How big does a black earth footprint have to be in order to be called ADE. Many archaeological occupation sites have discrete middens that meet the content criteria of ADE; but they are of small-scale contexts within the larger archaeological site (e.g. an individual garbage pit, lens of midden on an abandoned house floor, or post holes packed with dark midden). In addition to color, Kern et al. (2003) stress that ADE have a greater depth of anthropogenic A horizon than typical forest soils (30-60 cm vs 10-15 cm). Although the cultural strata of most archaeological sites correspond to the modern A Horizons; there are many exceptions such as those that are deeply buried paleosols or are so thick that post-abandonment soil formation processes have not created a deep A Horizon.

AMAZONIAN DARK EARTHS AND COMMUNITY SETTLEMENT PATTERNS

Archaeologists have become increasingly aware of the importance of cultural behavior and concepts regarding trash and its patterned, non-random disposal (e.g. Schiffer, 1987). As patterned trash disposal, ADE may provide insights into how people conceptualized and assigned cultural meaning to garbage (trash) and its proper disposal. Native Amazonians are known in the historical and ethnographic literature as "clean" people. Traditional villages are commonly described as clean, well maintained, and orderly. In the Upper Amazon, native peoples carefully sweep debris on and around house floors and house clearing to the outer edges of the village or hamlet daily. The cleared area of settlement has rich symbolic meaning as culturally domesticated social space distinguished from the wild, undomesticated space of the forest beyond. A clean village is part of community pride and great effects are made to prepare for important feasts. Clearings provide protection of bare feet against cuts by broken pottery and lithics and snakebite and for removal of decaying organic materials that harbor disease pathogens. Formal garbage disposal was critical for maintaining health in the large populated urban centers of late prehistory.

ADE are highly variable in size, form and depth of deposit. A working assumption held by most of the authors in this volume is that the size and depth of ADE are directly associated with population size of the settlement and settlement duration.8 Another assumption, although less discussed, is that the shape or "footprint" of ADE is associated with community pattern that reflects underlying social organization. Large, multicomponent (occupied for long periods of time by groups of people defined by distinct pottery styles and traditions) sites are found throughout the Amazon Basin regardless of whether ADE is present. Thus large size and long occupation duration are necessary, but not sufficient, explanations for ADE formation. In the following section, we explore two ideas: 1) presence and absence of ADE due to differences in settlement type and garbage disposal patterns, and 2) forms (spatial footprints) and internal heterogeneity of ADE reflect community settlement patterns and/or processes of settlement establishment and reestablishment of settlement space through time.

Could ADE patterns and forms be associated with a certain type of settlement design or village plan? Myers (1973) classified historical and ethnographic accounts of traditional community patterns into "linear" and "non-linear". Linear (one axis considerably longer than the other) includes "lines of houses community" and "multifamily longhouse community". While these basic categories of community pattern account for all ethnographic settlements or archaeological sites, Myers' study presents many valuable testable models.

The "lines of houses community", composed of end-to-end houses parallel to a long plaza overlooking a lake or river, is common in the Napo and Ucayali Rivers of the Peruvian and Ecuadorian Amazon. Garbage is disposed of in middens behind the houses and/or in front of the plaza. Myers (1973) attributes the pattern to the exigencies of a linear high ground in the form of narrow levees or bluffs along rivers. The large towns of continuous band of houses for kilometers along the Central and Lower Amazon River reported by early explorers were large versions of the lines of houses community. Apparently many settlements had houses arranged on streets facing public plazas with temples and men's houses. Formal roads leading to the interior are also described. The accounts are not specific about trash disposal patterns but one would assume that the pattern would be wide linear midden(s). The "multifamily longhouse community" or, made up of a single or multiple longhouses (maloca) with up to 100 families (400 people) per house is common in the northwest Amazon (Myers, 1973). The "multifamily long house community" of the northwest Amazon is linear because the domestic structure is much longer than wide (100 m long x 15 m wide; rectangular, oval, or combination in footprint). Garbage is tossed outside the house clearing along the axis of the longhouse.

Myers (1973) provides archaeological examples of linear ADE from the Upper Amazon that may have been formed by linear communities of lines of houses or single long houses. As Smith (1980) points out, the linear ADE in the Central and Lower Amazon extend hundreds of meters back from the bluff edge; thus, these communities must have had multiple rows of lines of houses. Many of these settlements may have been associated with the historical Tapajós chiefdom and archaeological ADE of a predominately linear type. Nimuendajú (1952) reports 65 Tapajós sites most of which are ADE and estimates that the total number is probably double.

Myers' non-linear pattern (circular, oval or amorphous in footprint) ranges from the "isolated single family house community" to the large "central plaza type community". I also add an intermediate type: the "multifamily roundhouse community" found in the Orinoco River drainage (Wilbert, 1981) and the "house lot community" based (discussed below). The "isolated single family house community" and the "house lot community" are common throughout Amazônia today. A single extended family house is surrounded by a 30-m diameter cleared area with a shallow "doughnut-shaped" ring of refuse around the clearing created through daily sweeping. Another hypothetical disposal pattern would be accumulation of garbage into a single heap. Refuse also accumulates under the house on the house floor, especially near cooking hearths.

A larger scale version of the above is the "multifamily roundhouse community" in which all reside within a single structure of 15 m diameter (Wilbert, 1981). The pattern of garbage disposal is expected to form a large doughnut shaped ring around the house and public plaza cleared area.

"Multiple longhouse communities" of the northwest Amazon and "multiple roundhouse communities" of northern South America are made up of various multifamily houses that together produce non-linear settlement pattern. Some of the larger communities are arranged around plazas. Garbage is either swept into a ring beyond the clearing of the cluster of long houses or in rings around each individual house. I would also add the "house lot community" based on traditional Maya urbanism (Killian, 1992). These are characterized by relatively regularly dispersed independent households with patterned spatial organization of house, outdoor activity areas, midden, and gardens.

The "central plaza-type community" is the most highly structured, non-linear community (Myers, 1973). Those described for the Bororo, Gê, and Carib-speaking peoples in Central Brazil are made up of a circular ring (or concentric rings) of up to 8-31 houses around a circular plaza of 110-300 m diameter. Communities of over 140 houses for up to 1600 people arranged in 3 rings around a plaza are documented. Garbage is placed in individual piles up to 10 m "behind" the houses that face the plaza. Over time, the piles of midden form a doughnut shaped ring beyond the house circle creating what would be a 350 m diameter or 10 ha site (Myers, 1973).

Myers predicts that the sweeping effect in the plaza and the mounding effect in the midden would create slight topographic differences. Variations include a central plaza-type community with a square or rectangular plaza and a street like arrangement of 4-8 multifamily longhouses each holding 30-200 families described for the Tupinamba of the Brazilian coast village of 4-8 houses, each housing 30-200 families (Lowie, 1948: 16). Colonial documents for Bolivia report large towns and villages of thousands of inhabitants with central plazas with up to 400 houses on streets, presumably organized as a rectangular or square grid (Denevan, 1966). The distribution of midden would probably be similar to that of the central plaza-type community or the house lot community. Myers (1973) notes that central plaza-type communities would be much easier to defend using ditches and palisades than linear villages.

Most small, dispersed shallow lenses of ADE on terra firme are probably isolated single family house communities but with a center type ADE instead of the doughnut shaped ring. The oval and elliptical mound ADE of the

Marajoara culture on Marajó Island (Roosevelt, 1991) and the Bolivian Amazon probably represent single or multiple long house communities (Myers, 1973). Archaeologically, the central plaza-type community includes the large central plaza sites often with arcs, circles, or rectangles of ditches and embankments described for the Upper Xingú Basin, Upper Madeira Basin (Miller, 1992, 1999), the Bolivian Amazon, the Açutuba and Osvaldo sites on the Rio Negro (Heckenberger et al., 1999), and ring plaza villages without earthworks for the Tocantins River Basin (Wüst 1994; Baretto and Wüst, 1999).

The oldest central plaza-type community archaeological sites are those of the Valdivia Culture in Ecuador (3500-1500 BC) with a large rectangular plaza surrounded by a ring of densely clustered multifamily houses.10 The majority of the central plaza-type community archaeological sites discussed above are concentric type ADE, in some cases with multiple rings of ADE such as on the Tocantins River (Wüst and Barreto, 1999). The primary settlement of the late prehistoric Tapajós culture is under the present day city of Santarém. The ADE is up to 1.5 m deep and estimated to cover 4-5 km2 and assumed to be non-linear.

Although both linear and non-linear community patterning are documented, Smith (1980) and Denevan (2001: 105) stress that most known ADE are linear (one axis longer than the other) and laid out parallel to the bluff edge and/or nearest active or once active river channel. The long axis of many of the larger ADE extends several kilometers ; thus, we may be seeing the accumulated result of many communities established and reestablished along the bluff over thousands of years. Another distinction raised in this volume is "center type" and "concentric type" ADE that may also reflect community patterning and complex site formation processes. Sombroek et al. (2003) characterize the concentric type ADE as having a deep concentric ring(s) of ADE around a relatively clean central area (the classic doughnut shaped midden discussed above) and the center type ADE in which the deepest anthropogenic soil is in the center and tapers off towards the edges of the site (mound-like midden). The concentric type ADE (e.g. best documented archaeologically in the Tocantins and Upper Xingú river basins by Wüst and Barreto, 1999) is assumed to correspond to the single house community and multiple family roundhouse community if small, and the central plaza community and its variants if large.

The majority of ADE are center type: they have continuous distribution of black earth across the site and are deepest in the center with no evidence of an ADE-free central plazas (Sombroek et al., 2003). The center type ADE does not closely fit any of the community patterns discussed above. Does this mean that pre-Columbian community was of a form that is not represented in the ethnographic and historical literature or could it be the result of complex site formation processes? The center type ADE implies that 1) people lived on top of their garbage (e.g. the moundbuilder model where trash and fill are used to raise the settlement; possibly for visibility, drainage, or an expression of monumentality); 2) garbage disposal for fields was spatially discrete from residence (e.g. Kayapó enriched garden model); 3) each community maintained a single location to pile garbage (rather than sweep it into a doughnut shaped ring around the residential clearing); and/or 4) slight shifts of the community location or residences within the community over time distributed the midden across the entire site.

Sombroek et al. (2003) and Denevan (2001) suggest that ADE may have started off as a concentric type but became center type due to a "smearing effect" caused by periodic reestablishment of the community through slight movement of the center of the settlement. Scholars have documented this phenomenon for Xinguano, Gê, and Bororo communities. In most cases, the community is moved several hundred meters because of "rotting houses, frequent deaths, internal disputes, warfare, and sanitary conditions" (Wüst and Barreto, 1999: 12). These hypotheses could be tested archaeologically with careful horizontal areal excavations and sampling of larger sites and ethnographic contexts. Archaeologists have shown in ethnographic and archaeological cases that despite careful daily sweeping, some garbage remains on house floors due to the "trampling effect" ; thus, the house clearing could accumulate continuous ADE given enough time. The buildup of midden on house floors and under storage racks over several years can be quite substantial. Few studies of differentiation of cultural space and heterogeneity of ADE through mapping and excavation have been done. Nimuendajú (1952: 11) pointed out, "The surface of the black earth deposits is usually not flat but composed of a number of mounds, each several meters in diameter, and each probably representing a house site".

In this case, the uneven surface of ADE may reflect the differential use of cultural space within the site. Heckenberger et al. (1999) detected a large rectangular plaza surrounded by artificial temple and/or elite residential mounds through surface topography, ditches, thickness of ADE, and distribution of artifacts at the Açutuba site. In most ethnographic cases discussed above, garbage middens and domestic zones within the settlement are spatially discrete (garbage is swept or tossed beyond the house clearing or central plaza house circle clearing). If all organic matter and charcoal generated by the community were systematically gathered to create ADE outside the residential zone in piles, ring, or arc of midden, or directly in fields, there should be a large, relatively organic-free zone in the residential sector of each site, no matter what type of community pattern. Center type ADE probably forms where sites have been continuously occupied for long periods of time. In large pre-Columbian settlements, it may not have been possible to dispose garbage far from the residence because of close neighbors; thus midden may have accumulated around and under houses as predicted in the house lot community or multiple reestablishments of any community pattern in the same general location.

Later inhabitants of the settlement would have to periodically decide whether to continue piling up new trash on the already established piles or rings inherited from their ancestors (maintaining a concentric type ADE) or to level the surface by filling in low spots with new trash (creating a center type ADE). In my experience digging archaeological sites in North and South America, organic matter, potsherds, and construction debris were used to level low spots, fill unused pits, and abandoned structures producing more dispersed distribution of localized ADE (creating greasy, black soils with considerable charcoal, ash, bone, and shell, what archaeologists usually referred to as "dark midden").

Most Amazonian peoples raise their house floors for improved drainage. Once organic matter has decomposed into mature ADE and not considered a health and aesthetic problem, it can be lived on or treated as transportable construction fill and fertilizer for agriculture. In this scenario, people literally lived on their garbage but only after it had been converted into harmless ADE. The presence of the house garden may have been a determining factor in formation of the center type ADE. Many authors of this volume (Clement et al., 2003; Hecht, 2003; Hiraoka et al., 2003; Silva, 2003) discuss the importance of the house garden as the prototype for the formation of ADE. The gardens that regularly receive organic matter from food preparation and cooking are located near the kitchen (thus residence and garden spaces are in close proximity).

In small settlements the garden is always adjoining the domestic space. As settlement size is increased, gardens can either be relocated outside the residential area or squeezed into spaces between houses (house lot community). There were fewer domestic animals (such as introduced pigs, goats, and chickens) to compete for house garden space in pre-Columbian settlements, thus house gardens may have been larger and more common than in modern villages. Over time, this strategy would generate a center type ADE.

CURRENT APPROACHES IN ECOLOGICAL ANTHROPOLOGY

There are a number of approaches to a human ecology that have been applied since the early 1980s. These represent the increasing specialization in anthropology, not only by the subfields that were described earlier on, but also by different theoretical approaches. Some approaches, parallel those taken in the field of ecology, but with time lags of several years.

Political Ecology

Political ecology is derived from political economy, in which there is concern with social inequalities and power relationships. It developed as a reaction to what some considered as an emphasis on ecological explanation for human social behavior to the neglect of political factors. The label has also been used in a Marxist context with the argument that "...an expanding capitalist economy is destructive to the environment." (Vayda and Walters, 1999). In this latter context, the ideas are principally Marxist in origin, not anthropological. Some anthropologists have criticized the contemporary application of political ecology as moving toward too much emphasis on "political" and too little emphasis on "ecological" relationships (Vayda and Walters, 1999).

Evolutionary Ecology

Evolutionary ecology arose from Mac Arthur's work in the 1960s that combined ideas from Darwinian evolution, ethology, population biology, and mathematical modeling. Much of the work deals with mathematical models of behavior within an adaptation framework. Anthropologists have been interested in this area of combined economic and ecological modeling of human behavior since the 1970s (Dyson-Hudson and Smith, 1978; Smith, 1979; Thomas et al., 1979). Sometimes the research is identified as behavioral ecology (Borgerhoff Mulder and Sellen,1994). Four areas of research that are germane to anthropology were identified:

(1) foraging strategies;
(2) mating systems and life-history strategies;
(3) spatial organization and group formation; and
(4) niche theory, population dynamics, and community structure (Smith, 1983).

Much of the research to date has focused on optimal foraging among hunter-gatherers such as the Peruvian Amazonian Piro (Alvard,1995), Ituri Pygmies (Bailey, 1991), Paraguayan Ache (Hill, 1988), Canadian Inujjuamiut (Smith, 1991), and Canadian Cree (Winterhalder, 1983). Other applications, such as optimal foraging of nomadic pastoralists have only been applied in a handful of cases. Borgerhoff Mulder and Sellen (1994: 225) identify the future of pastoralist studies as lying in "...a successful combination of quantitatively based studies and powerful modeling techniques," especially in the application of optimality models. Anthropologists are particularly well-suited to this kind of detailed observational research, because of lengthy time requirements for field observation within the tradition of extended field work in anthropology.

Historical Ecology

Historical ecologyis a relatively new approach in ecological anthropology that has been embraced by some archaeologists and ethnohistorians. Early works of interest to anthropologists were, among others, by Wiliam McNeill (1976) and Alfred Crosby (1972, 1986), historians who documented events linking human health and the environment in historical perspective. Based on a conference held at the New School of American Research in Santa Fe, New Mexico (Crumley, 1994), the field appears to be defined as a frame-work for studies of past ecosystems and their changes through time, with attempts to sort out the effects of anthropogenic and natural (non-anthropogenic) processes.

Most practitioners of this approach are from ethnohistory and archaeology, and they build their theory on ideas from landscape ecology, geography, archaeology, history, and ethnohistory.

This is an important application of ecology to anthropology, since losses in biodiversity during the present century can be placed in the context of earlier times through studies of prehistory. This will be discussed below in the context of biodiversity.

Landscape Ecology

Landscape ecology, with its background in geography and geomorphology, has a particular appeal to sociocultural anthropologists because of their current interests in land use in the Third World (Coppolillo, 2000). Archaeologists, as noted, are also drawn to this framework (in the context of historical ecology) for research because of the anthropogenic transformations of the landscape that are a part of human prehistory and history (Balée, 1998).

Ecosystems Ecology

Ecosystems ecology became the dominant research paradigm of the International Biological Programme (IBP), but still, the incorporation of human populations was limited (Worthington, 1975; Collins and Weiner, 1977). One of the problems with the Human Adaptability Component of the IBP was a conflict that arose between sociocultural and biological anthropologists during the early planning of the IBP.

Margaret Mead argued for a largely social approach to the Human Adaptability research, but was voted down at an ICSU General Assembly. She then pulled out her support for the program and the anthropological community followed suit (Weiner 1977). This then led to a dominance of the Human Adaptability research by human biologists and biological anthropologists.

Later, two conferences were held in an attempt to coordinate some of the IBP biome research with human adaptability research projects, but with limited success (Little and Friedman, 1973; Jamison et al., 1976). There is still resistance among some biological scientists who have interests in ecosystems analysis to incorporate human biologists or social scientists in their research for some of the obvious reasons discussed above.

However, more projects in the 1980s and 1990s, including those under the aegis of Unesco's Man and the Biosphere Programme, have successfully conducted collaborative ecosystems work on human populations.

Collaborative efforts between ecologists, other natural scientists, and social scientists have been initiated in the urban LTER (Long Term Ecological Research) projects. The Central Arizona-Phoenix LTER was codirected by an archaeologist, and the program is dependent on social scientists to develop realistic models within an ecosystems framework.

Ecology of Health and Adaptability

Studies of the ecology of health and adaptabilityof non-Western populations provide a breadth of environmental and health conditions not usually experienced by Western peoples.

It is therefore important to study traditional as well as industrial peoples to gain insights into the full spectrum of environmental influences on health. Ecological and biogeographical approaches and models are often useful in understanding health threats and risks. One model that has been useful employs the movement of people from one environment to another to test for effects of the new environment on health.

An IBP project investigated the effects of movement of nearly 1000 Pacific Tokelauan Islanders to New Zealand after a disastrous hurricane struck their island in 1966. Baseline health data of Tokelauans collected in 1963 were compared with New Zealand Tokelauan health, and these migrants were found to have higher prevalences of obesity, type II diabetes, asthma, and hypertension than the native islanders Studies of migrant American Samoans to Hawaii demonstrated essentially the same effects of migration and modernization. The principal variables associated with declines in health have been identified as diet, activity, and levels of stress associated with a Western life style.

Ecology of Reproduction

The Ecology of reproduction is an area of interest that is central to human population ecology and bears on the ecological processes that influence human reproduction. Biological scientists have known for a long time that the environment, particularly the availability of resources, profoundly influences reproduction. The earliest interest in this area was by anthropological demographers, with training in biobehavioral sciences, who were willing to entertain the possibility that humans were subject to the same biological rules as other organisms (Ellison 1990).

The advice of my close colleague in social anthropology, Neville Dyson-Hudson, who had a deep interest in livestock. During the planning of the Turkana research, we agreed that if we could identify environmental effects on the livestock, then we should look for similar effects in the human population. This was a radical view and would have been scorned by our sociocultural colleagues, but turned out to be a very productive way to generate hypotheses. It was particularly useful in Turkana studies of child growth, lactation and breastfeeding, maternal health, reproduction, and fertility.

2

Ecology and Environment

INTRODUCTION

Environmental conservation is an integral part of the socio economic development. The growing population, high degree of mechanisation and steep rise in energy use has led to activities that directly or indirectly affect the sustainability of the environment. The various kinds of pollution/pollutants may be broadly categorized as (a) Air pollution, (b) Water pollution, (c) Solid waste and (d) Industrial and hazardous waste. The sustainable use of bio-diversity is fundamental to ecologically sustainable development. India is one of the 12 mega diversity countries of the world.

The survival and well being of any nation depends on sustainable social and economic progress that satisfies the needs and aspiration of the present without compromising the interest of future generations. Environmental problems are largely the by-products of affluence marked by resource wasteful life style. Recycling of solid and liquid wastes, bio-composting, tree planting, etc. are important for environmental conservation.

MICROBIAL ECOLOGY—THE SCIENCE

Microbial ecology is a long-standing scientific discipline that is undergoing remarkable, even revo-lutionary, changes. The core of the field aims to understand microbial communities, which are self-organizing and self-sustaining assemblages of dif-ferentmicroorganisms,andhowthesecommunities interact with their environment.

The science of microbial ecology tries to an-swer four fundamental questions. First, what mi-croorganisms are present in the community? The phylogenetic makeup (identity and number) of the microorganisms present in a community is called a community structure. Next, what are the capabili-tiesofthemicroorganismsforcarryingoutreactions that transform the community's environment? The catalog of catalytic capabilities is called the com-munity's phenotypic potential. Third, what reactions and transformations are the community members actually performing? The realization of phenotypic potential is called the community's function. Final-and is still the most widely used. This rRNA gives information on the phylogenetic identity of the microorganisms, and this addresses the first fun-damental question of microbial ecology.

Other targets within the genetic system of micro-organisms answer different questions. For example, amplifying and detecting specific genes in the genome of microorganisms in the community defines the phenotypic potential, and this addresses the second question about the capability of microorganisms. Messenger RNA (mRNA) reveals which genes are being expressed and, therefore, which functional proteins like lyare formed.This gives information on the expressed phenotypic potential, or the third question about the community's function.

Assessing RNA targets by using microscopic visualization, such as with fluorescence in situ hybridization (FISH), adds information on spatial organization. Recently, microbial ecologists have begun to clone and sequence large genome fragments from various microbial communities; therefore, it is now possible, at least for low-diversity communities, to reconstruct entire genomes of uncultured community members. In addition, microauto-radiography, coupled with FISH (i.e.,MAR-FISH) and stable-isotope probing, makes it possible to detect the function in the community, even when the researcher has no idea what genes or mRNA should be detected.

Environmental biotechnology-The service Environmental biotechnology manages microbial communities that provide services to society. Prominent and emerging services include removing contaminants from water, waste water, sludge, sediment, or soil; capturing valuable products from renewable resources (e.g., biomass), particularly energy carriers but also nutrients, metals, and water; sensing contaminants or pathogens in the environment or, perhaps, in humans; and protecting the public from dangerous exposure to pathogens.

These services are essential if modern human society is to be safe, sustainable, and secure. Microbiological communities, properly managed, can provide these services reliably, continuously, economically, and without creating other hazards. Environmental biotechnology is almost a century old, although this name is relatively new. Past names include biological treatment, biological processes, bioprocess engineering, bioremediation, and bioenvironmental systems. The new name reflects that environmental biotechnology is adapting and benefiting from the modern tools of molecular biology, as well as from other advances in science and technology. The scientific foundation of environmental biotechnology is microbial ecology. Because environ-mental biotechnology ultimately aims to manage microbial communities for the good of society, a deep understanding of microbial communities-that is, microbial ecology is essential. Environmental biotechnology addresses real-world goals by managing microbial communities, rather than finding or creating the solves-all-problems "superbug".

The box at left lists criteria for a successful en-vironmental biotechnology for removing contami-nants or resource capture. These criteria demand a robust, self-organizing, and self-sustaining microbial community; they do not cater to a highly specialized or pampered superbug. Thus, the goal is to understand microbial communities so that the system's conditions lead to robust communities that provide the desired service.

Needs Pull, Science Pushes

A field that has been around for a century usually does not show a burst of creativity, but environ-mental biotechnology is doing just that today. The reason is the convergence of a strong "needs pull" with a strong "science push". Today, environmental biotechnology gets a powerful pull from the needs of human society arising from increasing challenges to achieve sustainability, environmental quality, se-curity, and human health.

The sustainability of modern human society depends on extracting essential materials from renewable resources and reducing reliance on non-renewable resources. At the top of the list of at-risk resources are two that environmental biotechnology addresses directly: water and energy. Today, water providers are forced to tap sources of lower quality:

polluted ground and river water, eutrophic lake and reservoir water, and wastewater. These poor-qual-ity sources need substantial treatment to eliminate public-health risks, unpleasant taste and odor, and discoloration. Experts on the Middle East predict thatwater-notoil-willbethegreatestfuturecause of strife in this region.

After water, energy is the most precious resource, and future sources must shift from fossil fuels to renewable ones. Environmental biotechnology is at the heart of upgrading poor water sources for human use and for converting renewable energy sources-particularly biomass and sunlight-to useful forms, including natural gas, hydrogen, and electricity. Organics, nutrients, and metals that are not captured but are instead discharged to the environment become pollutants, not resources. Environmental biotechnology has a long-standing role in treating wastewater and other contaminated water, air, and solids. Increasing population, urbanization, and economic activity heighten the need to apply environmental biotechnology to preserve (or improve) environmental quality, along with capturing valuable resources. Innovative environmental biotechnology approaches appear to be well suited for improving environmental quality in developed and developing countries alike. The readiness of developing countries to use these technologies lies in their improved reliability and operability in a decentralized setting.

Criteria for a Successful Environmental Biotechnology for Detoxification or Resource Capture

The technology must be practical at a large scale. rel-evant units normally are thousands of cubic meters per day for flow or tons for mass.

It must operate reliably and continuously around the clock. The concept of a "time-out" normally is not relevant. The technology must be relatively simple to operate and largely self-correcting without much hu-man intervention. It must be economical to build and operate. The services provided are so essential to society that they cannot be priced out of the reach of those on the low-est rung of society's economic ladder. Security and human health also relate to envi-ronmental biotechnology. Infectious diseases from pathogenic microorganisms remain the main cause of death worldwide, and terrorist threats to dissemi-nate pathogens heighten the danger. Microbial sys-tems can monitor for pathogens or chemical toxins in hospitals, water supplies, the air, and, perhaps, even humans. Microbial systems also contribute to cures or therapies by producing drugs or enzymes to fight diseases.

Environmental biotechnology can contribute in all these ways because microbial communities have seemingly infinite ways to live, even in environments that appear to be bizarre or hostile. Their ability to organize and sustain themselves provides human society with a cornucopia of metabolic po-tential to find services to benefit society. Microbial ecology, the core scientific discipline, allows us to understand microorganisms as part of their communities:"to think like the microorganisms." Armed with this deep insight, the environmental microbi-ologist can create sustainable systems that "work for the microorganisms so that they work for us." On the science side, the great advances in mi-crobial ecology, previously summarized, stand at the head of the science push and tend to garner the greatest credit. However, microbial ecology is not the only strong science push behind environmental biotechnology at the beginning of the 21st century. The products of modern materials science provide another hearty push, perhaps as important as from microbial ecology. The materials push began around 1970, when lightweight, high-strength plastics made possible biological towers for wastewater treatment, which were the first biofilm processes with high surface areas and small footprints.

From the 1980s through the 1990s, lightweight bio-film carriers in the form of gravel-sized pellets made even more compact, high-rate processes possible. From the late 1990s to today, microfiltration membranes have been replacing gravity separators for activated sludge, improving effluent quality, re-liability, and compactness. More recently, the membrane biofilm reactor makes it possible to use H_2 toreduceNO_3-,ClO_4-,andalargerangeofoxidized contaminants in drinking water, groundwater, and wastewater. Recent advancements in nanoscale (1-100-nm) materials and biomicro-electromechanical systems(bioMEMS) technologies surely will provide similar opportunities.

Another science push comes from mathematical modeling, the ultimate tool for integrating the large number of microbiological, chemical, and physical processes that occur in any microbial community of environmental significance. A model uses mass-balance equations to represent the significant com-ponents in the community. Creating a model demands that the modeler identify the important system components, for example, the critical types of microorganisms, the substrates they consume, and the products they produce. Likewise, the mod-eler must represent the important reactions, such as the synthesis of new biomass, consumption of substrates, and generation of products, with math-ematical expressions that capture what is known about the microorganisms and how they function.

On the one hand, modeling is an exercise in integrating all that we know about microorganisms so that we can test and exploit our understanding of how the microbial communities work as a functioning ecosystem.

On the other hand, modeling is the best way to find out what we do not understand about the microorganisms and their interrelationships with each other and their environment. When we cannot model a system, we can identify what we do not understand, and this helps us direct fundamental research to the most important questions.

Whether the goal is gaining scientific under-standing or applying that understanding to create high-value services or products, the most fundamental questions are at the interface of microbial ecology and environmental biotechnology.

Perhaps the overarching question is, "How do microbial communities self-assemble to achieve and maintain a function?" The environmental biotechnologist depends on self-assembly once the proper conditions are in place. Therefore, self-assembly needs to be based on identifiable principles that can be put into prac-tice.

The overarching question then leads to corollary questions: What are the underlying principles of self-assembly? What community structure is optimal for a biotechnological application? What environmental conditions trigger assembly of the desired community with its desired function? Are the conditions and outcomes predictable, reproducible, and controllable? Self-assembly also implies that microbial communities behave as a kind of multicellular organism, one in which the whole is more than the sum of its parts. Corollary questions along this line of inquiry include: Does the community have a multicellular A field that has been around for a century usually does not show a burst of creativity, but environmental biotechnology is doing just that today.

ECOLOGY MOVEMENTS AND CONFLICTS OVER NATURAL RESOURCES

The recent period in human history contrasts with all the earlier ones in its strikingly high rate of resource utilization. Ever expanding and intensifying industrial and agricultural production has generated increasing demands on the world's total stock and flow of resources. These demands are mostly generated from the industrially advanced countries of the North and the industrial enclaves in the underdeveloped countries of the South. Paradoxically, the increasing dependence of the industrialised societies on natural resources, through the rapid spread of energy and resource-intensive production technologies, has been accompanied by the spread of the myth that increased dependence on modern technologies implies a decreased dependence on nature and natural resources This myth is supported by the introduction of a long and indirect chain of resource utilisation which leaves invisible the real material resource demands of the industrial processes.

Through this combination of resource intensity at the material level and resource indifference at the conceptual and political levels, conflicts over natural resources generated by the new pattern of resource utilisation are generally shrouded and overlooked. These conflicts become visible when resource and energy-intensive industrial technologies are challenged by communities whose survival depends on the conservation of resources threatened by destruction and overexploitation, or when the devastatingly destructive potential of some industrial technologies is demonstrated as in the Bhopal disaster.

For centuries, vital natural resources like land, water and forests had been controlled and used collectively by village communities thus ensuring a sustainable use of these renewable resources. The first radical change in resource control and the emergence of major conflicts over natural resources induced by non-local factors was associated with colonial domination of this part of the world. Colonial domination systematically transformed the common vital resources into commodities for generating profits and growth of revenues. The first industrial revolution was to a large extent supported by this transformation of commons into commodities which permitted European industries access to the resources of South Asia.

With the collapse of the international colonial structure and the establishment of sovereign countries in the region, this international conflict over natural resources was expected to be reduced and replaced by resource policies guided by comprehensive national interests. However, resource use policies continued along the colonial pattern and, in the recent past, a second drastic change in resource use has been initiated to meet the international requirements and the demands of the elites in the Third World, leading to yet another acute conflict among the diverse interests. The most seriously threatened interest, in this conflict, appears to be that of the politically weak and socially disorganised group whose resource requirements are minimal and whose survival is primarily dependent directly on the products of nature outside the market system. Recent changes in resource utilisation have almost wholly by-passed the survival needs of these groups. These changes are primarily guided by the requirements of the countries of the North and of the elites of the South. This book analyses environmental conflicts in contemporary human society. In general it relates to societies all over the world, but in particular it addresses the most intense and emerging social contradictions in India related to conflicts over natural resources. Science and technology are central to these conflicts because while scientific knowledge has been used by contemporary societies to considerably enlarge man's access to natural resources, it has also allowed the utilisation natural resources at extremely high rates.

Emergence of Ecology Movements

The contemporary period is characterised by the emergence of ecology movements in all parts of the world which are attempting to redesign the pattern and extent of natural resource utilisation to ensure social equality and ecological sustainability. Ecology movements emerging from conflicts over natural resources and the people's right to survival are spreading in regions like the Indian subcontinent where most natural resources are already being utilised to fulfil the basic survival needs of a large majority of people.

The introduction of resource and energy-intensive production technologies under such conditions leads to economic growth for a small minority while, at the same time, undermines the material basis for the survival of the large majority.

In this way, ecology movements have questioned the validity of the dominant concepts and indicators of economic development. The ideology of economic development, which remained almost monolithic in the post World War II period, is thus faced with a major foundational challenge.

In this chapter an attempt has been made to provide a systematic conceptual framework for analysing the processes and structures of modern economic development from an ecological -perspective. It attempts to analyse the relationship between economic development and conflicts over natural resources to trace the roots of ecological movements. Further, in the light of the ecological perspective, it examines the fundamental assumptions and categories of modern development economics that are used to determine the objectives of economic development as well as the criteria for the choice of technologies that are used to achieve these objectives.

Ecology Movements and Survival

The intensity and range of ecology movements in independent India have continuously widened as predatory exploitation of natural resources to feed the process of development has increased in extent and intensity. This process has been characterised by the massive expansion of energy and resource-intensive industrial activity and major development projects like large dams, forest exploitation, mining and energy-intensive agriculture. The resource demand of development has led to the narrowing of the natural resource base for the survival of the economically poor and powerless, either by direct transfer of resources away from basic needs or by destruction of the essential ecological process that ensure renewability of the life-supporting natural resources.

In the light of this background, ecology movements emerged as the people's response to this new threat to their survival and as a demand for the ecological conservation of vital life-support systems. The most significant life-support systems in addition to clean air are the common property resources of water, forests and land on which the majority of the poor people of India depend for survival. It is the threat to these resources that has been the focus of ecology movements in the last few decades. Among the various ecology movements in India, the Chipko movement (embrace the trees to oppose fellings) is the most well known.

It began as a movement of the hill people in the state of Uttar Pradesh to save the forest resources from exploitation by contractors from outside.' It later evolved into an ecological movement that was aimed at the maintenance of the ecological stability of the major upland watersheds in India. Spontaneous people's response to save vital forest resources was seen in Jharkhand area in Bihar-Orissa border region as well as in Bastar area of Madbya Pradesh where there were attempts to convert the mixed natural forests into plantations of commercial tree species, to the complete detriment of the tribal people. In the southern part of India the Appiko movement, which was inspired by the success of the Chipko movement in the Himalayas, is actively involved in stopping illegal over-felling of forests and in replanting forest lands with multipurpose broad leaved tree species. In Himachal Pradesh the Chipko activists have intensified their opposition to the expansion of monoculture plantation of the commercial Chir Pine (Pinus roxburghii). In the Aravalli Hills of Rajasthan there has been a massive programme of tree planting to give employment to those hands which were hitherto engaged in felling of trees.

The exploitation of mineral resources, in particular the opencast mining in the sensitive watersheds of the Himalayas, the Western Ghats and Central India have also resulted in a great deal of environmental damage. As a consequence, environmental movements have come up in these regions to oppose the reckless mining operations. Most successful among them is the movement against limestone quarrying in the Doon Valley. Here, volunteers of the Chipko movement have led thousands of villagers, in peaceful resistance, to oppose the reckless functioning of limestone quarries that is seen by the people as a direct threat to their economic and physical survival.'

While the Doon Valley instance has a long history of popular opposition to the quarrying of limestone and a Supreme Court order has restricted the area of quarrying to a minimum, examples of such success' of ecology movements are rare People's ecology movements against mineral exploitation in the neighbouring areas of Almora and Pithoragarh still seem to be ignored, probably due to the relative isolation of these interior areas. Beyond the Himalayas, the ecology movement in the Gandhamardan Hills in Orissa against the ecological havoc of bauxite mining has gained momentum and it draws inspiration from the Chipko movement.

The mining project of the Bharat Aluminium Company (BALCO) in the Gandhamardan Hills is being opposed by local youth organisations and tribal people whose survival is directly under threat. The peaceful demonstrators have claimed that the project could be only continued 'over our dead bodies. The situation is more or less the same in large parts of Orissa-Madhya Pradesh region where rich mineral and coal deposits are being opened up for exploitation and thousands of people in these interior areas are being pushed to deprivation and destitution. This is also true of the coal mining areas around the energy capital of the country in Singrauli. In these interior areas of Central India, movements against both mining and forestry are becoming increasingly volatile and people's resistance is growing.

Large river valley projects, which are coming up in India at a very rapid pace, is another group of development projects against which people have organised ecology movements. The large-scale submersion of forest and agricultural lands, a prerequisite for the large river valley projects, always takes a heavy toll of dense forests and the best food growing lands. These have usually been the material basis for the survival of a large number of people in India, specially tribal people.

The Silent Valley project in Kerala was opposed by the ecology movement on the ground of its being a threat, not to the survival of the people directly, but to the gene pool of the Tropical Rainforests threatened by submersion. The ecological movement against the Tehri high dam in the UP Himalaya exposes the possible threat to people living both above and below the dam site through large-scale destabilization of land by seepage and strong seismic movements that could be induced by impoundment. The Tehri Dam Opposition Committee has appealed to the Supreme Court against the proposed dam by identifying it as a threat to the survival of all people living near the river Ganga up to West Bengal.

Most notable among the people's movements against dams on the issue of direct threat to survival from submersion are Bedthi lcchampalli, Bhopalpatnam, Narmada Sagar, Koel-Karo, Bodhghat, etc. In the context of the already overutilised land resources, the proper rehabilitation on a land-to-land basis of millions of people displaced through the construction of dams seems impossible. The cash compensation given instead is inadequate in all respects for providing an alternate livelihood for the majority of the displaced. Destitution is thus the first and foremost precondition for initiating large dam projects

Process of Construction of Dams

While the process of construction of dams itself invites opposition from ecology movements, the functioning of water projects dependent on the constructed dams results in further ecological disasters and movements. People's movements against widespread water-logging, salinisation and the resulting desertification in the command areas of many dams have been registered. Among them are instances of protests against the Tawa, Kosi, Gandak, Tungabhadra, Malaprabha, Ghatprabha projects and the canal irrigated areas of Punjab and Haryana. While excess water led to ecological destruction in these cases, improper and unsustainable use of water in the arid and semi-arid regions generated ecology movements in a different way. The anti-drought and desertification movement is gaining momentum in the dry areas of Maharashtra, Karnataka, Rajasthan, Orissa, etc. Ecological water use for survival is being advocated by water based movements like Pani Chetana, Pani Panchayat, and Mukti Sangharsh. Another major movement originating from the ecological destruction of resources by growth based development is spreading all along the 7,000 km long coastline of India. It is the movement of the small fishing communities against the ecological destruction caused by mechanised fishing whose instant profit motive is destroying the coastal ecology and its long-term biological productivity in a big way.

No amount of threat to survival in India from environmental hazards can be complete without a reference to the Bhopal tragedy on 2 December 1984, in which several thousand people died and several lakhs faced serious health hazards following the leakage of poisonous Methyl Iso Cyanate from a pesticide plant of Union Carbide (India) Limited. People's movements for clean air and water are growing in ail parts of the country just as ecologically irresponsible industrialization is moving deeper into the hinterland in search of new resources.

ECOLOGICAL CHARACTERISTICS TIDAL GRADIENT

In general, coasts with a greater tidal range can help in colonization of extensive mangrove forests. A greater tidal range increases the intertidal area and depending on the slope of the substrate, which if smooth encourages the growth of mangroves.

The topography of a mangrove swamp is normally smooth and flat. However, any rise or fall of the land can greatly influence the direction and rate of flow of the water in the system, which changes the colonization of mangroves.

Many species of mangroves respond differently to different tidal regimes. For example, in the Sundarban area of India, a mangrove forest that experiences total diurnal inundation is dominated by Avicennia marina and A. alba while Excoecaria agallocha, Ceriops dacandra and Acanthus ilicifolius get dominated at sites that are not completely inundated. Nypa fruticans also seems to prefer sites with low level of tidal inundation. If diversity is high, mangroves get extremely variable in their location. Their "zones" may be obscured by a broad overlap in species distribution or they may simply remain absent in some mangals.

Temperature

Mangroves attain climax growth only under tropical conditions where atmospheric temperature in the coldest months is greater than 20°C and the seasonal fluctuation does not exceed 5°C. Mangroves have been reported to grow in latitudes where the average sea surface temperature is 24°C.

Any further rise in temperature may lead to spreading of only some species in to higher latitudes, provided that the direction of ocean currents facilitates the dispersal of their seeds. However, very high temperatures are not favourable as leaves of mangroves are sensitive to temperature and their photosynthetic capacity gets reduced, falling to zero at leaf temperatures of 38-40°C, as against the optimum leaf temperature for photosynthesis which is 28-32°C. A small increase in temperature may not adversely affect the flowering, but may change their reproductive cycle and thus may alter the duration between flowering and the fall of ripe seeds.

Global warming may lead to sea level rise and as a consequence of that, mangroves may shift more landward. However, in many parts of the world, mangroves are unlikely to migrate landward because of human settlement at the landward boundary and the width of the mangrove forests is likely to decrease under the rise in sea level. The effects of sea level rise on any mangrove habitat will be influenced by local conditions such as the wetland type, geomorphic setting, and human activities in the wetland. It is predicted that the sea level rise may range from 45 to 65 cm/100 year. However, mangroves could survive a rise rate of 8-9 cm/100 year and hence the predicted faster rates could seriously threaten mangrove ecosystems. This view was challenged by Snedaker et al. (1994) who cited historical records showing mangrove expansion under relative sea level changes nearly twice as great. Tan and Zhang (1997) conclude that, given the predicted rates of estimate, the sea level rise poses no serious threat to most mangrove forests in China.

Salinity

Salinity plays a vital role in the distribution of species, their productivity and growth of mangrove forests. Changes in salinity are normally controlled by climate, hydrology, rainfall, topography and tidal flooding.

Mangroves generally tolerate higher salinity than do non-mangrove plants, but tolerance also varies among the mangroves. For example, Rhizophora mucronata seedlings do better in salinities of 30%, but R. apiculata do better at 15%. Sonneratia alba grows in waters between 2% and 18%, but S. lanceolata only tolerates salinities up to 2%.

In general, mangrove vegetation is more luxuriant in lower salinities. Experimental evidences indicate that at high salinity, mangroves spend more energy to maintain water balance and ion concentration rather than for primary production and growth. On the Pacific coast of Central America, freshwater availability (largely from rainfall and land runoff) controls reproductive phenology, growth and mortality of Avicennia bicolor. However, low salinity, associated with long periods of flooding, contributes to mangrove degradation through reduced cell turgidity and decreased respiration. Mangroves are poor competitors under non-saline areas where freshwater marsh plants easily out-class them. To cite an example, in Americas, the riverine mangrove forests disappeared at the Amazon and Orinoco river estuaries, where freshwater flow dominated over the tidal prism and under this situation; freshwater macrophytes replaced mangroves.

Salinity at high levels also affects mangroves. For instance, high salinity reduces the biomass in hydroponically grown Bruguiera gymnorrhiza, and causes denaturing of terminal buds in Rhizophora mangle seedlings. Saline interstitial water is known to reduce the leaf area, increases the osmotic pressure leaf sap, increases the leaf area/weight ratio, and decreases the total N, K, and P minerals. Simple salinity fluctuations also have significant negative effects on the photosynthesis and growth of plants. In Senegal, hypersalinity (from a decade of low rainfall and high evaporation) has caused the growth of salt marshes invading into mangrove areas and thus completely destroying their vegetation. Extremely high salt concentrations in the groundwater of tropical salt flats are also responsible for the complete absence of mangroves in such locations. There are often very sharp changes in the groundwater salt concentrations at the interface between salt flats and mangroves indicating that the mangroves are modifying the salinity of the groundwater.

Mathematical models developed for the groundwater flow show that human activities at hundreds of kilometers inland can destroy vast areas of mangroves by changing the flow of groundwater and thus modifying the salinity levels.

PUBLIC PARTICIPATION IN ENVIRONMENTAL DECISION MAKING

The title of this study, poses several questions: What is the 'public'? What is 'participation'? What are the 'decisions' to which it refers? Starting with the last and working backwards, EDM refers to any process of decision-making where consequent significant environmental impacts are a possibility. This includes law making, planning, strategic planning, resource management planning, licensing of industry e.g. IPPC, environmental assessment (EIA), spatial planning etc. EDM can be even more complex than decision-making on other public issues.

First, environmental impacts do not respect property, jurisdiction or boundaries.

Second, EDM can involve government agencies as both manager and regulator.

Thirdly, environmental issues can provide especially heated value conflicts that require value trade offs.

The level at which the public is involved varies with the relevant legislation, and the attitude of the other stakeholders. Often it just means informing the public of a previously made decision and asking for comments, which may or may not be heeded. Sometimes it means informed consultation. Here there is exchange of information prior to the relevant authority's reasoned decision making, and all inputs are included and seen to be included. An example of this being the River Basin Management process under Article 14 of the Water Framework Directive as described by Judith Cuff. Another example would be the Bantry Bay Coastal Zone Charter. Public participation, at its apex, could also mean that the public itself, in consultation with the relevant bodies, makes the final decision itself. Examples of this can be found under the auspices of Local Agenda 21. For public participation to be effective at any level, it requires the public to be well informed and kept aware of the possibility of participation. This requires a pro-active approach from industry and the relevant public bodies.

What then is 'the public'? The public is often treated as a unitary body, whereas in reality it is a collection of numerous continually shifting interests and alliances (Ortolano), which may be in conflict with each other. The term is used as a "catch-all to describe those with an interest in a decision, other than a proponent, operator, or responsible authority". (Petts and Leach). The individuals making up a public may be involved as individuals or as members of organisations. They may become involved due to their proximity, economics, social or environmental issues, values, etc. By contrast, stakeholders, of which the public is one, are literally those with a stake in an issue and may include non-governmental organizations (NGO's), government or its agents, industry, individuals, communities etc.

Stakeholders do not always want to be involved in an EDM process, but they have the right to know if their interests are affected. They may want to become involved at a different stake of the EDM process.

Benefits, Costs and the Forces Affecting Public Participation in EDM

Public participation in EDM is playing an increasingly important role in the protection of the environment. The forces driving the increased use of participatory practice, the benefits accruing from same.

Forces Driving Stakeholder Processes

Many different forces are at work, prior to, and throughout, a public participation programme. However, the following factors have also moved the participation debate forward. People generally resist change, especially when they do not understand or agree with the goals, the methods, the sponsor or the timing of the proposed change. So, keeping the public in the dark is often a recipe for disaster. Citizens are also increasingly reluctant to defer to 'expert' government agency opinions, and are unwilling to act as sounding boards for bodies that have already made decisions, particularly when they affect their local communities.

Both of these tendencies lead to a lack of public confidence and trust, unless the relevant authorities take note of them. There are growing social expectations for environmental equality resulting in part from the high media profile given to environmental issues. Findings, that minority and low income groups in the USA had been disproportionately affected by pollution and downgraded environments, resulted in the issuing of Executive Order 12898 by President Clinton.

A far-reaching programme ensued, based on meaningful involvement and fair treatment for all races, cultures and incomes in environmental decision-making. "The transmittance of 'experience', rather than knowledge, is the critical neglected dimension of decision-making, particularly as it pertains to our national decision-making". For example development planning in Dublin City requires largely the same expertise as rural development planning, but the experience, vision and values of the rural population may be very different to that of their city cousins. The further the EDM process is from central government, the more pertinent becomes the public participation in same, and in this context, face-to-face meetings are critical to producing real learning and trust.

Previously, with the exception of Development Plans and Waste Management Plans, public involvement in EDM in Ireland was largely restricted to the minimal fulfillment of the two "rules of national and constitutional justice, and in particular 'audi alteram partem' (hear both sides)".

In practice this meant that firstly a decision was made and this was then presented to the public for comment, a process also known as 'report and comment'. So, both sides have been heard, but one has the power, and can choose to ignore the input of the other. Changes driven by the Aarhus Declaration and Principal 10 will affect many areas of environmental regulation and planning, opening them up to far greater public participation. Alongside this is the increasing transparency of regulatory processes.

The directive on Freedom of Access to Information on the Environment (Directive 90/313/EC) SI.125 of 1998 and The Freedom of Information Act, 1997 have opened up the EDM processes to closer scrutiny by the public. By compelling the relevant authorities to release documents on request, the inner workings of government have been opened up for closer examination, and relevant information freed up to enable more effective public participation in environmental governance. There is also a growing diffusion of information. The many portals of information available on the internet, together with the traditional media, libraries and the excellent work, in Ireland, of ENFO, all increase awareness of environmental issues.

Since the Aarhus Convention, which agreed to recognise and support the crucial role played by ENGO's in articulating opinions of an environmentally concerned public, European funding has been given to support their activities. €4.62 million was distributed in Europe in 2003. In Ireland, the Department of the Environment and Local Government allocated €132,000 in 2001, and €133,000 in 2002 to build capacity in 23 ENGOs with a national remit.

This was done via a limited company known as Environmental (Ecological) Core Funding Ltd.. It was estimated in 1999 that there were some 500 International ENGOs (Environmental NGOs) world-wide that operated in three or more countries. Apart from the rapid growth of the ENGOs, there is clearly expanded interest in participation. Public unrest at the prospect of development, without formal opportunity for participation in decision-making, is increasingly responsible for expensive delays. In the UK, policing of antiroad protests has alone cost millions of pounds (Treweek, J.1999). The Shell Oil Company, in the UK, was eventually forced to adopt a major consultation programme on the fate of the Brent Spar, North Sea oil exploration rig, as a result of the international reaction to the campaign waged by Greenpeace..

Hampshire County Council, in England, with a population of some 2,000,000, had to scrap its waste management plan, following much public unrest, and start from scratch, with a well planned and successful public participation programme. A waste plan was then adopted which had broad support, and is being put into action.. Many industry organizations are realizing that for their good image, public safety, and the longterm financial well-being of their members, responsible partnership with host communities is essential.

There is some evidence that corporations which build sound, responsive relationships with all their stakeholders tend to be longer lived and more successful as businesses in the long term than other corporations (Wheeler and Sillanpaa, 1997). Olson, M. and Toyne, P.(2000) also report a study which suggests a direct correlation between increased public involvement of the public in corporate affairs, and increased profits.

The Chemical Manufacturers Association in the U.S.A. (representing 90% of all chemical production) set up the Responsible Care Initiative, which sets codes of practice. There are six areas covered by these codes viz: Community Awareness and Emergency Response (CAER); The Pollution Code; The Process Safety Code; The Distribution Code; The Employee Health and Safety Code; and The Product Stewardship Code. In relation to all of these, the guiding principles exhort their members "to recognize and respond to community concerns about chemicals and our operations, and to work with others to solve problems created by past handling and disposal of hazardous substances." The adoption, by some industries, of environmental management systems, such as I.S.O. 14001, and E.M.A.S (Eco-Management and Audit Systems) shows their belief that a transparent and responsive industry is also more likely to be a viable one. E.M.A.S. requires public reporting of environmental management and auditing. Internationally, trends towards decentralization in many countries, increases in the numbers of countries with representative democratic forms of government, and conditions laid down by donor agencies on development loans, e.g. The World Bank, are some of the international forces bringing increased public involvement in EDM. Whilst it is not the intention of the author to go into detail on risk assessment and disaster management., it is worth noting that public participation should, under Council Directive 96/82/EC (Seveso ?) and the Council Decision 1999/847/EC, play an important part in both of these important areas. Wates, N. (2000) gives guidelines for organizing community risk assessment and disaster management programmes, whilst the "UK Guidelines for Risk Assessment and Management" (Defra UK. 2002) give a good introductory insight into the social aspects of risk, and in particular the factors influencing risk perceptions. These factors are listed here and are also relevant to other areas of EDM.

- o Risks which are involuntarily imposed (e.g. pollution from an incinerator) tend to be seen as less acceptable than voluntary ones (e.g. bonfires)
- o Unfamiliar risks (e.g. GMO's) tend to cause concern especially when it appears that they are poorly understood by science.
- o Activities associated with a dreaded form of death (e.g. cancer) are viewed with alarm.
- o Man-made or technological risks (e.g. nuclear power) are less acceptable than natural ones (e.g. Radon). o A risk involving a single large-scale consequence (e.g. civil aviation accident) causes more concern than risks which result in numerous small scale consequences (e.g. car accidents).
- o Alarm may be caused by risks when the consequences of exposure are delayed and cause hidden or irreversible damage (e.g. ionizing radiations).
- o Inequitable distribution of risks and benefits as a result of a particular activity is likely to make the risk less acceptable. An example here might be the remaining operative Chernobyl nuclear power stations. Employees, knowing the risks involved but needing to earn a living, would have a different view of risk to others living in the vicinity, and different again to those living 100 kms away, and relying on the electricity to heat their homes in temperatures of-30°C.
- o Activities posing a threat to certain groups such as children or future generations are generally more worrying.
- o Risks which are the subject of controversy and contradictory information generally cause concern.

Benefits of Public Participation

The protection and enhancement of the environment is the main aim of most EDM processes, and public participation can improve the quality of both the process and the end decision. The participation process is also a learning one for all the stakeholders involved, especially if there is a free flow of information between the parties.

For the public it can often be the first experience of taking an active part in the democratic process. EDM processes benefit from the direct and immediate knowledge held by citizens and business, concerning environmental conditions in their communities and industries. Encouraging the public and other stakeholders to share their knowledge, with the regulatory authorities, fosters better-informed decisions and decreases the likelihood of environmental harm, whilst increasing project viability. Of 25 overseas projects sponsored and evaluated by the World Bank, 13 failed mainly through lack of local input (E.L.I. Research Report, 1997) (Technical Document 1, 1996). It should be realized that "wisdom is not limited to scientific specialists and government officials," and that "rational analysis, carried on in ignorance of political reality, may well end up so divorced form social reality, as to be of little use to anyone." (Barkenbus, J. 1998, p.2).

Public participation gives broader perspectives on a particular process, and early involvement gives: added time to study issues and develop the process; enhanced credibility of the decision making process; early identification of the diverse perspectives on the issues of concern; and the generation of solution options (E.L.I. Research Report, 1997). Better-designed projects, which avoid costly delays in appraisal and implementation, can also result from early and planned consultations and public participation. For the proponent, early participation can have the added benefit of diffusing opposition to a project. If a broad based consensus is built, it can also lead to a public sense of 'ownership'. The public's enthusiasm is potentially a powerful motivating force for the EDM process. Public input can also supplement scarce government monitoring, inspection, and enforcement, resources.

Involvement in EDM is a learning experience giving the public insight into the governance process. If the experience is good, it powers the way for future co-operation. The converse can also be true! Experience has also shown that environmental issues are powerful catalysts of civic 9 participation and serve as a good incentive for citizens' action and responsible democracy (Technical Document 1, 1996). Executive Order 12898 introduced by President Clinton is a far-reaching programme, based on meaningful involvement and fair treatment for all races, cultures and incomes in environmental decision-making. It has done much to prevent the dumping of dirty industry on the marginalized members of the USA citizenry. The order requires state agencies, and in particular the EPA, to pro-actively seek out the disenfranchised members of communities likely to be affected by developments, or already living in depleted environments, and encourage them to become involved in the relevant EDM processes.

Costs of Public Participation

There is limited information on the relative costs of using the different methods of participation practice, although some attempt is made to quantify, which describes each method. Actual costings for specific processes are given in a number of studies. Clearly deliberative processes, which are used to engage relevant stakeholders in debate, discussion and deliberation, could only be done as part of a major project. Expenses range from €20-40,000 (for a citizen jury with 16 participants for 5 days), to €200-300,000 (for 3 community advisory groups of 16 people over 6 months).

However, extensive use of participatory processes in the 'Wise Use of Flood Plains' project showed costs of about 10% of the above. Most of this funding was used to employ facilitators. The costs to members of the public in time and money, as well as emotional stress in situations where access to EDM is limited, do not appear to have been estimated previously. An attempt is made to gather some facts about the first two in this study. The intensity of the stakeholder process, including commitments of time, energy and money, and the often, uncertain results, has led to a burnout phenomenon amongst many participants, and from all stakeholder groups. Excessive raised expectations about participation can lead to frustration and anger with a breakdown in trust. It is essential then to be clear with all participants as to the limits of influence of the process. Where an infrastructure project crosses planning jurisdiction boundaries or where more than one regulator is involved, the costs to the public, state and proponents are multiplied. The increased time scale and expense involved in public participation is of particular significance in projects with a small capital funding. An international example of this would be the AIJ projects (Activities Implemented Jointly under the Climate Change Convention). These projects already require appraisal by two governments and then there is the extra of public participations costs. After this they must compete with traditional energy supply systems, where often no regulatory costs are involved.

Problems Encountered in Public Participation Practice

The main areas of difficulty affecting public participation are attitudinal, lack of capacity to deliver programmes, lack of clarity about what outcomes are possible and the lack of a legislative framework. Some aspects of these issues are addressed below. The difference between success and failure is not always clear. If the public participation process delivers a clear decision, which is then overruled by the legislature, is this a success? A good example here is the Bantry Bay Integrated Coastal Zone Management (ICZM) Project, funded for three years under the EU Life Programme. After three years of intense community activity and the drawing up and enactment of a text-book public participation plan, the Bantry Bay Coastal Zone Charter was created. At this point, no further funding was available, and the Project was closed down. Clearly the process was a success, but the result not. There is a general lack of experience of participatory processes in Ireland, and very few stakeholders have experienced planned participation processes. There may be suspicion, cynicism, or enthusiasm, but there is unlikely to be previous training. Suspicion about participation could be from politicians, who may feel their power is being diluted, or from ENGOs that have very specific and fixed viewpoints. Cynicism could be from previous public participation processes which were poorly managed, leaving participants reluctant to try again (E.L.I. Research Report, 1997). Citizens are frustrated when they are treated as adversaries, rather than welcome participants in the EDM process........ they feel they have been invited too late in the process. Public 'hearings' often do not include 'listening'. Citizens efforts and ideas are not included in proposals and no reason given.

The resources available to proponents and government are overwhelming. Litigation and direct action at least get a reaction". Similarly, public hearings are criticised for: being held at times inconvenient for the public, establishing an atmosphere that inhibits dialogue, and conducting proceedings that intimidate the public (Lein, K.L. 2003, p.56).

The lack of a clearly defined purpose can also means that the role of the public is sometimes unclear not only to the public but to those delivering the process, with the consequence that the participation process chosen is often unsuitable for the particular EDM process (Breggin, L and Hallman, H. 1999. p.44). The making available of the necessary resources for a worthwhile public participation process, whether by the proponent or the regulator, by the commitment of personnel and finances, is essential. The implications for project proponents, of not having a properly planned and funded participatory EDM process, may prove much more costly, than dealing with the backlash. In many circumstances, the choice increasingly is not whether to involve the public, but how to get the best value from the chosen process (Petts, J. and Leach, B. 2000).

To avoid hearing from only the activist or the powerful elite, and in order to get the widest sweep of opinion and information, authorities must reach out into the community. In this regard some activities are best avoided, for example public meetings.

In the absence of trained facilitators these can often come down to 'he who shouts loudest wins!' Lack of technical support for the public, and difficulties in getting access to information can diminish the ability of the public to play a meaningful part in EDM processes. Public access to independent expertise in a particular technical/scientific field can prove difficult, particularly where the proponent of a project is a monopoly employer in the industry in the state. Faced with five weeks to object to a planning proposal, for example, this is a major hurdle. Added to this, it is the experience of the author that stakeholders with no technical/scientific background find it problematic to access, comprehend or evaluate data and information, especially under these short time constraints.

The feelings of powerlessness engendered by these hurdles, add to public perceptions of a lack of influence at both lower and higher levels of national and European government. "Consultations of interested parties.... can only ever supplement, and never replace, procedures and decisions of legislative bodies which possess democratic legitimacy". This sense of powerlessness may be part of the reason why, even in cases where considerable energy and resources are expended, to identify individuals and groups, only a small proportion of the public ever attend participation programmes. Public involvement may actually result in an increased level of conflict. Becoming involved in a consultative process where the decision has already been made, or where the possible outcomes are not made clear at the outset, can lead to a great deal of frustration.

Anger may follow when the input of the public is ignored, especially following prolonged constructive engagement with an EDM process. Conflict can also occur between professionals. One source of such disagreements could evolve from the different emphases of 'scientific peer review' and 'social peer review'. The scientific peer review process is well established, and is essential to assess the technical information provided to policy makers. The parallel social peer review, designed to obtain societal acceptance and legitimacy for the decisions rendered, by contrast, does not have a set of recognised professional standards. It should also be remembered that whilst science may strive for the 'truth', the opinions of scientists are coloured by their values and beliefs. Scientists, engineers and other 'experts' should show respect in the face of 'emotional', cultural and traditional forms of wisdom. The perceptions created by a technocratic philosophy encourage NIMBY activism and further polarise communities.

The resulting deadlock frustrates the ability of planners to achieve consensus on initial issues and contributes to anti-participation attitudes. Therefore, rather than define 'NIMBY'attitudes as an irrational response to problems ordinary citizens cannot grasp, the solution is to encourage more participation not less. The misconception that the public are overly emotional, ill-informed and solely motivated by selfish interests, in their obstruction of facilities for the common good, leads to an information-based strategy. The assumption being that, 'if only the public knew the truth!' This assumes the existence of an objective truth to know. The conflict that arises from having different bodies making decisions on different parts of the same EDM process can often confuse the public understanding and increase frustrations. For example the executive split of planning and environmental matters in some aspects of development between the Local Authority and the EPA. Public involvement in these circumstances must deal with two different agencies.

Organisations resist change, and communication blocks within organisations can prevent them from making timely responses to their publics. The established routines and organizational systems of many bodies seek to promote the 'status quo'. Organisational structures which support sustainability, and therefore public participation, are a threat to 'command and control' style management systems. Managers, who have risen to power in this style of structure, frequently resist the transition to alternative structures, which embrace public consultation. Many opportunities for public participation are laid down in the environmental legal framework, as shown in Appendix 1. Where legislation does not specify early public involvement, then it is at the discretion of the relevant authority, and dependent on the opinion of the relevant public servant.

For example, in the development planning process, the Manager's reporting of the consultative process clearly includes the names and nature of all submissions together with his response to them, whereas in making an Air Quality Management Plan (under Section 48 of the Air Pollution Act, 1987 and Section 102 of the Environmental Protection Agency Act 1982) there is no legal compulsion to show how any submission might have affected the final decision.The recent use by Cork City Council of Article 81 of The Planning and Development Act, Regulations, 2001, to enter into a public/private partnership with a company previously refused planning permission to develop the same site (TP 02/26234), leaves the public only minimal scope for comment and no route to appeal any decision made.Under section 19 of the Protection Of The Environment Act, 2003, the making, variation or replacement of a Waste Management Plan, has been made an Executive Function. This has removed the immediate political accountability of the decision-making, and so removed another route for public input.

3

Water Environment and Pollution

WATER POLLUTION

Water pollution is a broad and generic term with a variety of meanings. *It can, in fact, mean almost any type of aquatic contamination between two extremes*:

1. A highly enriched overproductive biotic community, as a lake or river enriched with nitrates and phosphates from domestic sewage;
2. A biotic community with sufficient concentration of toxic substances to eliminate many forms of living organisms or even exclude all forms of life.

Types of water pollution may also be identified by the medium in which they occur (surface water, ground water, soil, etc.), the habitat in which they occur (marine, estuarine, river, etc.), or the source or type of contamination (nutrient, domestic, pesticide, thermal, industrial, etc.). One of the best definitions of pollution is.

"Environmental pollution is the unfavorable alteration of our surroundings, wholly or largely as a by-product of man's actions, through direct or indirect effects of changes in energy patterns, radiation levels, chemical and physical constitution and the abundance organisms." Pollutants of streams, lakes and estuaries come from many sources. Excessive nutrients commonly originate in domestic sewage and run-off from agricultural fertilizer. Certainly the former is the major source of excessive nutrients in most streams and lakes. Toxic chemicals originate in industrial operations, acid waters from mine seepage or surface erosion, and washings of herbicides and insecticides.

Bacterial and Viral Pollution

Water pollution becomes not only an esthetic problem for man, but an endonomic and medical one as well. Bacterial and viral contamination is a threat for the spread of water-borne diseases such as typhoid, shigellosis or bacillary dysentery, amoebic dysentery, cholera and hepatitis. In many watersheds raw sewage is a serious problem. The Hudson River above New York City receives over 200 million gallons per day of raw sewage.

Around many estuarine systems and freshwater lakes and even ocear beaches, public swimming areas have been closed in recent years because of high bacterial counts resulting from domestic pollution. During the summer of 1971, the tourist trade at beach resorts in Belgium, France, Spain and Italy was adversely affected by coastal pollution.

Approximately one-half the public bathing beaches of Chesapeake Bay within a twenty mile radius of Baltimore have been closed by public health authorities because of high coliform bacterial counts. These bacteria are taken as indicators of pollution, often, but not exclusively, emanating from domestic sewage.

In Florida, oyster beds in Tampa Bay, Pensacola Bay, Indian River and the St. John's River are closed to harvesting because of domestic pollution. Many parts of San Francisco Bay are closed to water contact sports, as are many beaches of Lake Erie near the larger towns and cities. This does not mean that all these communities live on the brink of typhoid or dysentery epidemics, though the potential does exist in times of flooding, but it does mean that human activities and recreation must be curtailed because of environmental deterioration.

There is still much to be learned about the bacterial communities of polluted aquatic systems. Normally we do not think of serious enteric pathogens being in our water systems in any great abundance. Coliform bacteria, the common indicators of domestic pollution, are not pathogenic in their common forms. The US Public Health Service estimated that a minimum of 40,000 cases of waterborne illness occur every year in the United States. As specific examples, in Riverside and Medera, California, in 1965, waterborne disease caused 20,000 illnesses and several deaths. Approximately 8 million Americans drink water with a bacteriological content exceeding the recommended standards of the USPHS.

Effects of Excessive Nutrients

Even with modern sewage treatment plants, water pollution problems are not entirely avoided. Modern plants remove or inactivate bacteria from the effluent water, but such water is still rich in basic nutrients, such as ammonia, nitrogen, nitrates, nitrites and phosphates. Primary and secondary sewage plants do not remove these sources of pollution. Primary sewage treatment involves screening and sedimentation of solids; secondary treatment involves biological reduction of organic matter; only tertiary treatment removes nutrients. Such nutrients stimulate plant growth, often in the form of phytoplankton or algae. The enriched waters are thus prone to plankton blooms which may have several undesirable consequences. Some plankton blooms, particularly those of the blue-green algae, produce undesirable odors and tastes in water. Others, such as the dinoflagellate blooms or redtide of the southern coastal regions, produce toxic metabolic products which can result in major fish kills.

Plankton blooms of green algae do not necessarily produce undesirable odors or toxic products, but they can still create problems of oxygen supply in the water. While these blooms exist under abundant sunlight, they contribute oxygen to the water through photosynthesis, but under conditions of continued cloudiness, they consume more oxygen than they produce and lead to oxygen depletion in the waters. Thus, dissolved oxygen may decline rapidly from favorable levels of 10 to 12 ppm to unfavorable levels of 2 to 3 ppm in which fish experience distress and asphyxiation. Highly enriched streams just below sewage outfalls may show a severe reduction in fish populations, as was documented for the Patuxent River of Maryland. The Potomac River below Washington, DC is highly polluted with domestic sewage and has dissolved oxygen levels often less than 1 ppm. This portion of the river displays annual fish kills every May, when fish reach these oxygen-depleted waters during their spring migrations.

Excessive nutrient levels in aquatic systems can cause two other kinds of ecologic consequences. They may lead to extensive growth of aquatic weeds such as Eurasian milfoil, water hyacinth, water chestnut, and many others which have become a worldwide problem. These growths may become so great as to impair fishing, bathing, fish spawning, shellfish production, and even navigation. Hence, excessive plant growths in enriched waters often represent a major economic problem as well as a complete disruption of aquatic ecology.

It has recently been demonstrated that excessive nutrients in water supplies, in the order of 8 or 9 ppm of nitrate nitrogen can cause human disease, as, for example, methemoglobinemia in infants. This is an illness caused by a modified form of normal oxyhemoglobin in the blood, resulting in inadequate oxygen transport by red cells and labored breathing.

Industrial Pollutants

The effects of industrial wastes on aquatic systems could be the subject of an entire book, and we can do no more here than provide a few examples. In Bellingham Bay and the Straits of Juan de Fuca north of Seattle, Washington, sulfite wastes from pulp mills produce abnormal growth of oyster larvae.

Sometimes the effects of pollution are much more dramatic than the modification of larval growth. In 1959, the US Public Health Service started a national survey of pollution-caused fish kills. In 1961, 45 states reported major fish kills, and in 1962, 38 states reported pollution-caused fish kills totaling 381 separate incident. Of these 381 fish kills, 43 per cent were attributed to industrial wastes, 13 per cent to agricultural poisons, 8 per cent to domestic sewage, 5 per cent to mining operations and 31 per cent were of unknown origin.

Some of the various conditions or accidents which caused these kills were as follows:

- Chemicals were used to flush the lines of a power plant (sodium nitrite, hydrazine, ammonium bifluoride, and soda ash);
- Bags of endrin used for insect control were accidentally dumped into a stream;
- The unauthorized application of copper sulfate was used to control a plankton bloom;
- Outfall from a paper mill contained lignite and sulfite waste liquor;
- Acid mine drainage discharged into a stream;
- Concentrated sulfuric acid released from a cotton seed delinting plant;
- Seepage of the pesticides chlordane and heptachlor from termite treatment project;
- Chromium solution from a plating tank leaked into a sewer system;
- Cyanide released from drains of blast furnaces at a steel company;
- Hot water from a steam generating plant released into a stream.

These are merely representative examples of the types of toxic effects which industrial and agricultural operations may have unless rigorous control measures are maintained. Increasing use of powerful chemicals in industrial and agricultural operations increases the risk of environmental damage through accidents.

Occasionally fish kills involve massive populations of aquatic life. In 1962 in San Diego harbor, an estimated 37,800,000 fish were killed by pollution, producing one raft of dead fish 1000 feet long, 10 feet wide and 3 feet deep. In 1963, an extensive fish kill in Chesapeake Bay killed millions of fish from Baltimore to Norfolk, and in 1967 many acres of dead alewife fish washed up along the Lake Michigan shores of the Chicago waterfront. The Chesapeake Bay and Lake Michigan fish kills may have been the result of epizootics of infectious disease, however, and were not necessarily the direct result of pollution. Certainly pollution played a role in these outbreaks, however, and may have been a triggering factor.

Another major problem in water pollution is the addition of various ions and chemicals in the water which have toxic effects on plant, animal and human life. Chlorine added to water to control bacteria and algae may persist in streams to cause mortality of plankton and fish. This occurred in Maryland in chlorine-treated waters used to cool an electric power plant on the Patuxent River. Mercury, as a by-product of industrial operations involved in the production of vinyl chloride, has cropped up as a toxic agent of serious proportions.

Mercury is used in many chemical industries, and it also emerges as a by-product of some incinerators, power plants, laboratories, and even hospitals, to the extent of 23 million pounds per year throughout the world. In Japan, human illness and death occurred in the 1950s among fishermen who ingested fish, crabs and shellfish contaminated with a simple alkyl mercury compound from Japanese coastal industries. This mercury poisoning produced a crippling and often fatal disease known as Minamata disease.

The name was derived from Minamata Bay on the southwest coast of Kyushu. Minamata disease was characterized by substantial pathology of the central nervous system and voluntary musculature. Initial symptoms included numbness of the limbs; lips and tongue, impairment of motor control, deafness, and blurring of vision. Cellular degeneration occurred in the cerebellum, midbrain, and cerebral cortex, and this led to spasticity, rigidity, stupor, and coma.

The first cases of Minamata disease appeared in Japan in 1953, and of 52 original patients in the villages around Minamata Bay, 17 died, 23 were permanently disabled, and only 3 were able to return to work within 6 months. Minamata disease has not occurred in the United States as a result of eating coastal fish, but in 1969 it was discovered that fish in Lake St. Clair and Lake Erie contained mercury above permissible levels. The Canadian government immediately banned the commercial sale of fish from these waters until the industrial sources of mercury were removed or corrected, and the mercury levels of fish fell below critical limits. In 1971, the US Food and Drug Administration banned the consumption of tinned swordfish because 80 per cent of the samples tested had excessive levels of mercury. Some samples of tuna fish have also shown high levels of mercury, and in 1971 a New York woman who had been dieting by eating tuna every day developed mercury poisoning.

Mercury is just one of many toxic chemicals which may occur in water polluted by industrial or agricultural effluents, Lead, cadmium, and nickel carbonyl are other pollutants which are toxic or pethogenic for man and animals. In fact, more than 12,000 toxic industrial chemicals are in use today. With over one-half of our total available fresh water row being used, one can glimpse the immense ecologic problem confronting modern society.

A basic concern in the overall picture of water pollution is the finite nature of our present surface and ground water supplies. Water use for domestic and industrial purposes is increasing so dramatically, that the time when all available fresh water will be used is now in sight for many communities. Hydrologists estimate the total fresh water supplies of the United States at 515 billion gallons per day, of which we are already using 360 billion gallons per day. The metropolitan areas of Philadelphia and Wilmington are now using nearly 80 per cent of their total available water. This shall reach 100 per cent within 10 years at present consumption trends. The great hope, of course, lies in finding new sources and in desalination, but these solutions contain economic and technological problems that remain to be solved. Water pollution and water supply are intimately related because of the necessity for and economics of reuse.

COMMON CAUSES OF WATER POLLUTION

Oil

Petroleum often pollutes water in the form of oil. Oil spills from ships and super- tankers, and from off-shore oil drilling operations cause pollution. Oil and petrol that leaks from cars and trucks also washes off roads and into waterways through storm water drains. Oil forms a thin layer on top of water and act like a lid on the surface and the water. Animals and plants living in the water can't breathe, the oil coats the feathers of water birds, and the fur of animals that swim in the water, causing them to become sick and, if there is a great amount of oil on their bodies, to die. Even the insects that live on the surface of the water are badly affected.

Fertilizers

Fertilizers contain nutrients such as nitrates and phosphates that help plants to grow. That's why farmers use them. When fertilizers are washed into rivers and streams the nitrates and phosphates cause excessive growth of water plants. The plants clogs the waterways, use up oxygen in the water, and block light to deeper waters. This is harmful to the fish and other invertebrates that live in water because it make it hard for the animals to breathe.

Soil

Pollution of waterways is also caused when silt and soil washes off ploughed fields, construction and logging sites, and from river banks when it rains.

Sewage and other Organic Pollutants

When material such as leaves and grass clippings, and waste from farm animals enters the water, it rots and breaks down and uses up the oxygen in the water. Many types of fish and other aquatic animals cannot survive.

Organisms such as bacteria and viruses enter waterways through untreated sewage in storm-water drains, run-off from septic tanks, and from boats whose owners dump sewage into the water. These microscopic pollutants cause sickness in people and in animals that drink or live in the water.

Chemicals

Chemical pollution entering rivers and streams causes great destruction. The chemicals can come from factories, construction sites, mining operations, and from homes when people pour chemicals down the sink or down the toilet.

Plastics

Floating plastic is ugly, and harmful to the environment. Plastic rubbish is not biodegradable (it doesn't rot away after we have used it) It can choke animals that try to eat it, and drown those that get tangled in it.

Litter

When people drop litter such as plastic and cans, food wrappers and cigarette butts, they can be washed by the rain into rivers and other waterways through storm water drains in the streets.

At the beach, it is important that people take home their litter or put it into garbage bins at the beach so that it doesn't get into the sea.

Other Causes of Water Pollution Include:

- Air Pollution
- Carbon Dioxide
- Fuel Additives
- Increased Water Temperature
- Industrial Development
- Landfills
- Mining Activities
- Gasoline
- Household Cleaning Products
- Personal Care Products
- Pesticides
- Pharmaceuticals
- Sediment

The Effects of Water Pollution

The effects of water pollution are far-reaching. And it's not only humans who are affected. All plants and animals must endure poisonous drinking water, river and lake ecosystems that have become unbalanced and can no longer support biodiversity. Deforestation from acid rain can occur as well. On the whole, water pollution has long-term effects on our health and economic productivity. Fight water pollution to keep our planet safe. The effects of water pollution differ from region to region, depending on the pollutants in the water and environmental factors. Common effects of water pollution include unhealthy or poisonous water, sick animals that pass their sickness on to humans, ecosystems that are unable to support a normal diverse animal and plant habitat and more.

Pollution Affects the Food Chain

Pollution in the form of organic material enters waterways in many different forms as sewage, as leaves and grass clippings, or as runoff from livestock feedlots and pastures. When natural bacteria and protozoan in the water break down this organic material, they destroyed the whole things.

Pollution Affect Aquatic Ecosystems

In nature nothing exists alone. Living things relate to each other as well as to their non-living, but supporting, environments.These complex relationships are called ecosystems. Each body of water is a delicately balanced ecosystem in continuous interaction with the surrounding air and land.

Whatever occurs on the land and in the air also affects the water. If a substance enters a river or lake, the water can purify itself biologically — but only to a degree. Whether it is in the smallest stream or lake — or even in the mighty oceans — the water can absorb only so much. It reaches a point where the natural cleaning processes can no longer cope. For example, various species of fish now suffer from tumors and lesions, and their reproductive capacities are decreasing

Pollution Affect Marine Life

Lakes many are persistent toxic chemicals such as DDT. Populations of fish consuming birds and mammals also seem to be on the decline. Of the ten most highly valued species of fish in Lake Ontario, seven have now almost totally vanished. People dump hazardous materials into the ocean to get rid of them. Sewage and waste from factories and cities can reach the ocean. This pollution is very harmful. It can kill the plants and animals living in the ocean. Dangerous chemicals like mercury kill the aquatic environment very quickly.

Water Pollution's Chain Reaction

One of the many causes of water pollution is sewage and fertilizers that contain nutrients such as nitrates and phosphates. When these enter our water system in excess levels, the growth of aquatic plants and algae is over-stimulated. As a result of the excessive growth of these aquatic plants, our waterways are clogged.

They use up dissolved oxygen as they decompose, and block light to deeper waters. Subsequently, the respiration ability or fish and other invertebrates that reside in water are also damaged.

EUTROPHICATION

Eutrophication is caused by the decrease of an ecosystem with chemical nutrients, typically compounds containing nitrogen or phosphorus.

It may occur on land or in the water. Eutrophication is frequently a result of nutrient pollution such as the release of sewage effluent into natural waters (rivers or coasts) although it may occur naturally in situations where nutrients accumulate (e.g. depositional environments) or where they flow into systems on an ephemeral basis (e.g. intermittent upwelling in coastal systems).

Eutrophication generally promotes excessive plant growth and decay, favors certain weedy species over others, and is likely to cause severe reductions in water quality. In aquatic environments, enhanced growth of choking aquatic vegetation or phytoplankton (that is, an algal bloom) disrupts normal functioning of the ecosystem, causing a variety of problems. Human society is impacted as well: eutrophication decreases the resource value of rivers, lakes, and estuaries such that recreation, fishing, hunting, and aesthetic enjoyment are hindered.

Health-related problems can occur where eutrophic conditions interfere with drinking water treatment. Although traditionally thought of as enrichment of aquatic systems by addition of fertilizers into lakes, bays, or other semi-enclosed waters (even slow-moving rivers), terrestrial ecosystems are subject to similarly adverse impacts. Increased content of nitrates in soil frequently leads to undesirable changes in vegetation composition and many plant species are endangered as a result of eutrophication in terrestric ecosystems, e.g. majority of orchid species in Europe.

Ecosystems (like some meadows, forests and bogs that are characterized by low nutrient content and species-rich, slowly growing vegetation adapted to lower nutrient levels) are overgrown by faster growing and more competitive species-poor vegetation, like tall grasses, that can take advantage of unnaturally elevated nitrogen level and the area may be changed beyond recognition and vulnerable species may be lost.

Eg. species-rich fens are overtaken by reed or reedgrass species, spectacular forest undergrowth affected by run-off from nearby fertilized field is turned into a thick nettle and bramble shrub.

Eutrophication was recognized as a pollution problem in European and North American lakes and reservoirs in the mid-20th century.

Since then, it has become more widespread. Surveys showed that 54% of lakes in Asia are eutrophic; in Europe, 53%; in North America, 48%; in South America, 41%; and in Africa, 28%.

Concept of Eutrophication

Eutrophication can be a natural process in lakes, as they fill in through geological time, though other lakes are known to demonstrate the reverse process, becoming less nutrient rich with time.

Estuaries also tend to be naturally eutrophic because land-derived nutrients are concentrated where run-off enters the marine environment in a confined channel and mixing of relatively high nutrient fresh water with low nutrient marine water occurs.

Phosphorus is often regarded as the main culprit in cases of eutrophication in lakes subjected to point source pollution from sewage. The concentration of algae and the tropic state of lakes correspond well to phosphorus levels in water. Studies conducted in the Experimental Lakes Area in Ontario have shown a relationship between the addition of phosphorus and the rate of eutrophication.

Humankind has increased the rate of phosphorus cycling on Earth by four times, mainly due to agricultural fertilizer production and application. Between 1950 and 1995, 600,000,000 tonnes of phosphorus were applied to Earth's surface, primarily on croplands (Carpenter et al. 1998). Control of point sources of phosphorus have resulted in rapid control of eutrophication, mainly due to policy changes.

Human activities can accelerate the rate at which nutrients enter ecosystems. Runoff from agriculture and development, pollution from septic systems and sewers, and other human-related activities increase the flux of both inorganic nutrients and organic substances into terrestrial, aquatic, and coastal marine ecosystems (including coral reefs). Elevated atmospheric compounds of nitrogen can increase soil nitrogen availability.

Chemical forms of nitrogen are most often of concern with regard to eutrophication because plants have high nitrogen requirements so that additions of nitrogen compounds stimulate plant growth (primary production). Nitrogen is not readily available in soil because N_2, a gaseous form of nitrogen, is very stable and unavailable directly to higher plants. Terrestrial ecosystems rely on microbial nitrogen fixation to convert N_2 into other physical forms (such as nitrates). However, there is a limit to how much nitrogen can be utilized. Ecosystems receiving more nitrogen than the plants require are called nitrogen-saturated. Saturated terrestrial ecosystems contribute both inorganic and organic nitrogen to freshwater, coastal, and marine eutrophication, where nitrogen is also typically a limiting nutrient. However, in marine environments, phosphorus may be limiting because it is leached from the soil at a much slower rate than nitrogen, which are highly insoluble.

Ecological Effects

Adverse effects of eutrophication on lakes, reservoirs, rivers and coastal marine waters (from Carpenter et al., 1998; modified from Smith 1998). Many ecological effects can arise from stimulating primary production, but there are three particularly troubling ecological impacts: decreased biodiversity, changes in species composition and dominance, and toxicity effects.

Decreased Biodiversity

When an ecosystem experiences an increase in nutrients, primary producers reap the benefits first. In aquatic ecosystems, species such as algae experience a population increase (called an algal bloom). Algal blooms limit the sunlight available to bottom-dwelling organisms and cause wide swings in the amount of dissolved oxygen in the water.

Oxygen is required by all respiring plants and animals and it is replenished in daylight by photosynthesizing plants and algae.

Under eutrophic conditions, dissolved oxygen greatly increases during the day, but is greatly reduced after dark by the respiring algae and by microorganisms that feed on the increasing mass of dead algae. When dissolved oxygen levels decline to hypoxic levels, fish and other marine animals suffocate. As a result, creatures such as fish, shrimp, and especially immobile bottom dwellers die off. In extreme cases, anaerobic conditions ensue, promoting growth of bacteria such as Clostridium botulinum that produces toxins deadly to birds and mammals. Zones where this occurs are known as dead zones.

New Species Invasion

Eutrophication may cause competitive release by making abundant a normally limiting nutrient. This process causes shifts in the species composition of ecosystems. For instance, an increase in nitrogen might allow new, competitive species to invade and outcompete original inhabitant species. This has been shown to occur in New England salt marshes.

Toxicity

Some algal blooms, otherwise called "nuisance algae" or "harmful algal blooms," are toxic to plants and animals. Toxic compounds they produce can make their way up the food chain, resulting in animal mortality. Freshwater algal blooms can pose a threat to livestock. When the algae die or are eaten, neuro-and hepatotoxins are released which can kill animals and may pose a threat to humans.

An example of algal toxins working their way into humans is the case of shellfish poisoning.

Biotoxins created during algal blooms are taken up by shellfish (mussels, oysters), leading to these human foods acquiring the toxicity and poisoning humans.

Examples include paralytic, neurotoxic, and diarrhoetic shellfish poisoning. Other marine animals can be vectors for such toxins, as in the case of ciguatera, where it is typically a predator fish that accumulates the toxin and then poisons humans.

Nitrogen can also cause toxic effects directly. When this nutrient is leached into groundwater, drinking water can be affected because concentrations of nitrogen are not filtered out. Nitrate (NO_3) has been shown to be toxic to human babies. This is because bacteria can live in their digestive tract that convert nitrate to nitrite (NO_2). Nitrite reacts with hemoglobin to form methemoglobin, a form that does not carry oxygen. The baby essentially suffocates as its body receives insufficient oxygen.

Sources of High Nutrient Runoff

Characteristics of point and nonpoint sources of chemical inputs (from Carpenter et al, 1998; modified from Novonty and Olem 1994)

In order to gauge how to best prevent eutrophication from occurring, specific sources that contribute to nutrient loading must be identified. There are two common sources of nutrients and organic matter: point and nonpoint sources.

Point Sources

Point sources are directly attributable to one influence. In point sources the nutrient waste travels directly from source to water. For example, factories that have waste discharge pipes directly leading into a water body would be classified as a point source. Point sources are relatively easy to regulate.

Nonpoint Sources

Nonpoint source pollution (also known as 'diffuse' or 'runoff' pollution) is that which comes from ill-defined and diffuse sources. Nonpoint sources are difficult to regulate and usually vary spatially and temporally (with season, precipitation, and other irregular events).

It has been shown that nitrogen transport is correlated with various indices of human activity in watersheds, including the amount of development. Agriculture and development are activities that contribute most to nutrient loading. There are three reasons that nonpoint sources are especially troublesome:

Soil Retention

Nutrients from human activities tend to accumulate in soils and remain there for years. It has been shown that the amount of phosphorus lost to surface waters increases linearly with the amount of phosphorus in the soil. Thus much of the nutrient loading in soil eventually makes its way to water. Nitrogen, similarly, has a turnover time of decades or more.

Runoff to Surface Water and Leaching to Groundwater

Nutrients from human activities tend to travel from land to either surface or ground water. Nitrogen in particular is removed through storm drains, sewage pipes, and other forms of surface runoff. Nutrient losses in runoff and leachate are often associated with agriculture. Modern agriculture often involves the application of nutrients onto fields in order to maximise production.

However, farmers frequently apply more nutrients than are taken up by crops or pastures. Regulations aimed at minimising nutrient exports from agriculture are typically far less stringent than those placed on sewage treatment plants and other point source polluters.

Atmospheric Deposition

Nitrogen is released into the air because of ammonia volatilization and nitrous oxide production. The combustion of fossil fuels is a large human-initiated contributor to atmospheric nitrogen pollution.

Atmospheric deposition (e.g., in the form of acid rain) can also effect nutrient concentration in water, especially in highly industrialized regions.

Other Causes

Any factor that causes increased nutrient concentrations can potentially lead to eutrophication.

In modeling eutrophication, the rate of water renewal plays a critical role; stagnant water is allowed to collect more nutrients than bodies with replenished water supplies.

It has also been shown that the drying of wetlands causes an increase in nutrient concentration and subsequent eutrophication booms.

Prevention and Reversal

Eutrophication poses a problem not only to ecosystems, but to humans as well. Reducing eutrophication should be a key concern when considering future policy, and a sustainable solution for everyone, including farmers and ranchers, seems feasible. While eutrophication does pose problems, humans should be aware that natural runoff (which causes algal blooms in the wild) is common in ecosystems and should thus not reverse nutrient concentrations beyond normal levels.

Effectiveness

Cleanup measures have been mostly, but not completely, successful. Finnish phosphorus removal measures started in the mid-1970s and have targeted rivers and lakes polluted by industrial and municipal discharges. These efforts have had a 90% removal efficiency. Still, some targeted point sources did not show a decrease in runoff despite reduction efforts.

Minimizing Nonpoint Pollution: Future Work

Nonpoint pollution is the most difficult source of nutrients to manage. The literature suggests, though, that when these sources are controlled, eutrophication decreases. The following steps are recommended to minimize the amount of pollution that can enter aquatic ecosystems from ambiguous sources.

Riparian Buffer Zones

Studies show that intercepting non-point pollution between the source and the water is a successful mean of prevention. Riparian buffer zones, an interface between a flowing body of water and land, have been created near waterways in an attempt to filter pollutants; sediment and nutrients are deposited here instead of in water. Creating buffer zones near farms and roads is another possible way to prevent nutrients from traveling too far. Still, studies have shown that the effects of atmospheric nitrogen pollution can reach far past the buffer zone. This suggests that the most effective means of prevention is from the primary source.

Prevention Policy

Laws regulating the discharge and treatment of sewage have led to dramatic nutrient reductions to surrounding ecosystems, but it is generally agreed that a policy regulating agricultural use of fertilizer and animal waste must be imposed. In Japan the amount of nitrogen produced by livestock is adequate to serve the fertilizer needs for the agriculture industry.

Thus, it is not unreasonable to command livestock owners to clean up animal waste-which when left stagnant will leach into ground water.

Nitrogen Testing and Modeling

Soil Nitrogen Testing (N-Testing) is a technique that helps farmers optimize the amount of fertilizer applied to crops. By testing fields with this method, farmers saw a decrease in fertilizer application costs, a decrease in nitrogen lost to surrounding sources, or both.

By testing the soil and modeling the bare minimum amount of fertilizer needed, farmers reap economic benefits while the environment remains clean.

Natural State of Algal Blooms

Although the intensity, frequency and extent of algal blooms has tended to increase in response to human activity and human-induced eutrophication, algal blooms are a naturally-occurring phenomenon.

The rise and fall of algae populations, as with the population of other living things, is a feature of a healthy ecosystem. Rectification actions aimed at abating eutrophication and algal blooms are usually desirable, but the focus of intervention should not necessarily be aimed at eliminating blooms, but towards creating a sustainable balance that maintains or improves ecosystem health.

GLOBAL ENVIRONMENTAL PROBLEMS AND LOCAL POVERTY

Poverty can be global. In terms of environmental research, there has been an increasing focus on the interrelationship between environment and development, and how nature and society affect each other through a complex interplay. This approach also implies that global environmental changes can have very different impacts on different local communities. Conservation of the baobab tree (Adansonia digitata) on cultivated land represents one climate adaptation in arid regions in East Africa. The fruit is a useful source of nutrition during drought years. The IPCC's suggestion that poverty plays an important role in determining how climate changes will affect various populations has two important consequences for policy design principles: The first is that those who are most vulnerable to the impacts of climate change should be the primary target of mitigation policy; poverty issues should be taken into account as a central element in climate measures. The second consequence is that policy design must take into account that the impacts of climate changes and possible effective measures to reduce vulnerability vary both from place to place and from time to time because social conditions change. It will be just as important to identify how adaptability can be improved among various social groups, as it will be to identify physical climate impacts. Measures should therefore be prepared in cooperation with the local population, where their priorities and conception of the problems become the centre of focus.

Single problem, even though the causes of the degradation are considered to be separate phenomena at a global level. Environmental problems that are addressed separately at an international level - such as climate, biodiversity, and decertification - must at be tackled holistically at the local level. There are several practical examples of the interaction between several "different" environmental issues in locally adapted measures. One study in two arid agricultural areas (one in the Kitui district in Kenya and the other in the Same district in Tanzania) shows that biodiversity in the form of local tree and plant species distributed throughout the farms represents an alternative income source for poor farmers when crops fail. For example, local wood is used to make stools, kitchen equipment, chicken coops, and so on. These products are sold on the local markets. Lumber is also used to burn charcoal, which is sold to urban areas and cities. Leaves and seeds from certain trees are used as feed for goats and cattle. Drought resistant indigenous fruits are an important source of nutrition for both children and adults when there is little food. Thus preservation of local knowledge and biodiversity in cultivated areas helps enhance adaptability and reduce vulnerability to extreme climate events such as drought and flooding. Preservation of the natural vegetation is also useful in combating desertification.

Moreover, particular ridges and hills covered with natural vegetation in dry areas have great value with respect to biodiversity. These areas often have different compositions of plant and animal species that otherwise are not found in the surrounding lowlands. These ridges, which have higher precipitation than the lowlands, form important parts of local watersheds. In addition, there is a better microclimate (local temperatures and moisture) around these forested areas, particularly in arid regions that are dependent on the forests as water sources. Where ridge vegetation is well preserved, forest streams will often be active also during the dry season.

A similar case of where climate and biodiversity measures have been coordinated is in the northern coast of Vietnam and described by Nguyen Hoang Tri and colleagues (1998). Preservation of the natural mangrove forest with its species diversity is important for local sources of income, for example, wood and honey production.

The mangrove forest is equally important for protection from cyclones and typhoons. The coast is hit by between one and twelve typhoons per year, and there is great uncertainty about how the frequency and magnitude of these will be changed in connection with global warming. The mangrove forest protects agriculture against flood damage from cyclones and reduces the maintenance costs of the dikes. Preservation of this forest represents a type of climate adaptation. These examples illustrate that the interaction between various "global" environmental problems can result in different impacts in different places. To reduce the vulnerability of the poorest populations, policy measures should be adapted to local conditions and address several different environmental problems at the same time. In developing countries seeking to expand their economic activities, consideration for environmental conservation often receives a low priority. In addition, approaches used in industrialized countries often cannot be applied directly in developing countries. In this context, NIES is conducting research on ways to conserve the environment that are appropriate for developing country conditions.

Water quality and air pollution are serious problems in developing countries in the Asian region. Air pollution in major cities marked by many factories and heavy vehicle traffic also has high concentrations of sulphur dioxide and suspended particulate matter (SPM), at levels Japan experienced in the past. In addition, problems such as damage from acid rain and transboundary pollution are growing more serious. Pollution of rivers and lakes from chemical substances (including agricultural chemicals) and eutrophication (including abnormal growth of toxic algae) are also occurring more frequently, while water shortages and tropical forest destruction are worsening.

While many developing countries give economic development the greatest priority, many problems remain with basic needs such as safe drinking water and food, as well as medical and public health services. This situation often hinders progress in addressing environmental problems. In some countries, including Bangladesh, China and India, negative health impacts are growing over large areas due to fluorine and arsenic pollution in air and drinking water implementing measures directed at developing countries. The Global Environment Facility (GEF) is the Climate Convention's funding scheme, designed to assist developing countries in developing climate strategies and integrating climate considerations into public policy, government, and development plans. The GEF is also supposed to fund preparation and implementation of practical measures, but until now no such GEF supported projects have been implemented. Most of the GEF's financial assistance has been directed at greenhouse gas emissions rather than enhancing local adaptation to climate impacts. As the poorest populations are responsible for very low greenhouse gas emissions but are nevertheless the most vulnerable to possible climate changes, it is thus a challenge to adapt the GEF to support measures designed to assist the poor. To truly integrate poverty considerations - which, for example, can be connected to local property rights - into environmental measures is also a great challenge because the GEF is essentially not a development fund but rather an environment fund. The GEF was established to fund measures specially directed at biodiversity and climate, and development measures are defined as outside its mandate.

CLEAN DEVELOPMENT MECHANISM

The Kyoto Protocol's Clean Development Mechanism (CDM) is another important mechanism to finance measures in developing countries. This mechanism is designed to enable industrialized countries to meet their commitments to the Kyoto Protocol by financing development projects that reduce emissions of greenhouse gases in a developing country. Negotiations are still ongoing with respect to the regulatory framework of the CDM, but pilot projects have already been implemented. It is important that the CDM provide feasible solutions that benefit poor developing countries and that the CDM projects are not designed on the premises of the industrialized countries in a top-down fashion. Another important source of funding for environmental measures in developing countries is development aid organizations such as NORAD or the World Bank. Since climate issues have previously only minimally been seen as an important development issue, only a few such organizations have paid attention to climate impacts. It will therefore be a challenge both to increase the awareness of the climate issue among these organizations and to enhance the cooperation between these organizations and the UN system in the design of policy measures. This is particularly important because development organizations have valuable experience with how locally adapted measures can be implemented to reach the poorest populations. The focus on poverty and local development is an important step forward in combating the negative impacts of climate change, but at the same time presents tremendous practical challenges. The question is how meaningful local solutions can be found to global problems.

For a long time man has assumed that the environment enveloping his existence was in vulnerable, bountiful, and immense in its capacity to support life. Events of the last two centuries have belied these presuppositions. The exuberance of the new era of technology and abundance has evaporated and the bright prospects of economic progress and superabundant living have been shattered by the discovery of the finiteness of the planet. With all the technological developments, the world still suffers from the scourges of poverty, disease, and misery, along with environmental degradation, destruction of forests, and extinction of many species.

The basic fact of this planet is that it is overpopulated. An increasing number of people are becoming urbanized, they continue to consume resources for survival, and to utilize them at an accelerated pace. By the end of the century, the population will swell to about 6.5 billion, two-thirds of whom will be living in developing countries. The prospect of supporting such a vast human population on the planet seems stupendous and grim, making the future of man unpredictable. It is now clearly recognized that urban man is continuously destroying many useful elements of life by using them indiscriminately and, consequently, degrading the environment which sustains life. It has been scientifically established that the ecosystem of the Spaceship Earth is fragile, limited, and that man's tampering with it may, in the end, make the earth unlivable, not only for man but for all life forms. Man has begun to live in large agglomerations of millions of people; he has command of technologies, which can take him to the farthest planets in the solar system; and he has unravelled mysteries that were secrets of nature only 100 years ago. In a sense, the twentieth century could have been a turning point in man's consciousness about himself and his universe, which should have freed the human race from the perennial threats of destitution, privation, and death. In reality, however, urban life in many parts of the world has become an insult to mankind because of the total degradation of people and their environment. In this context, UNESCO's "Man and the Biosphere Program" (MAB) states its objective to be "development of a basis through natural and social sciences for the rational use and conservation of the resources of the biosphere and improvement of the relation between man and the environment." Further, it states that "the consequences of today's actions on tomorrow's world should be predicted and man must manage natural resources efficiently." This report succinctly underscores the uncertain future of man if the symbiotic relationship between man and his environment continues to deteriorate at the present rate.

All through the two million years of man's existence on the planet, his survival has been conditioned by the intricate balance among the various elements of the earth, which have provided a continuous supply of materials for his needs. During the last 300 years, how ever, man's relationship with the earth, its environment, and outer space has changed drastically be cause of the accumulation of vast knowledge, increasing technology, exponential increase in expectations and needs, and, finally, a quantum jump in human activities by exploding population.

The technological revolution has ushered in a new era of transformation, which has not left any part of, the biosphere untouched, directly or indirectly, making man's survival increasingly precarious. Urbanization, as well, has resulted in a new order of relationships between human society and the ecosystem. The Earth went through a series of drastic changes for billions of years until, ultimately, life originated, leading eventually to the evolution of man. Only recently, the complex web of human activities has generated extensive modifications of the environment. Every moment man is displacing millions of tons of various elements from one area to another, transforming them from one form to another, and converting them into a large number of products through different processes. For example, 800 million tons of various metals are taken from the earth, 40 million tons of toxic chemicals, and 7 billion tons of conditional fuel units are consumed every year.

In short, the immense number of industrial activities of man is producing vast quantities of pollutants, which result in massive degradation of the environment. An alarming threat to the life-preserving ecosystem is evident around the globe. It can be observed in acid rains, disappearance of plant and animal species, desertification, smog, destruction of coastal areas, dirty waterways, poisoned foods, and blaring noises. Looking at the plight of the enduring earth, what Barry Commoner once said seems to be true: "We have been living under a vast and potentially fatal illusion: that we can enjoy the enormous benefits of modern technology with out risk to the integrity of human life and the environment."

Slowly, land, air, and water are being vandalized by man's vast, indiscriminate plundering of the environment. How far the ecosystem of the earth will be able to bear the threat before collapsing is only a question of time. Exploding population, increasingly complex technology, and careless use of resources have joined to face us with a triple danger. Unless man finds another suitable environment or creates a new one, or unless he adapts to the increasingly polluted environment, he must put the planet in order by restoring natural equilibrium if he does not want to face extinction.

HOW TO REDUCE WATER POLLUTION

1. Reduce the amount of runoff that comes from your property. Reducing runoff pollution actually has two components: improving the quality of runoff and reducing the quantity.
2. Maintain your vehicle. You can see the stains from leaky cars all over any parking lot. The chemicals—motor oil, transmission fluid, and antifreeze, just to name a few—almost always get washed directly into the nearest river or body of water. Have your vehicle regularly serviced and immediately repair any leaks you notice. Driving less or getting rid of your car entirely will do a tremendous service to the environment.
3. Minimize your use of fertilizers, pesticides, and herbicides. The chemicals you spray or spread on your home, lawn, or garden don't stay there. Traces of these poisons get washed into storm drains with rainwater or snowmelt. Multiply these small amounts by thousands of households, and the effects on watersheds and aquatic life can be catastrophic. Think twice before using these products, and consider alternatives (i.e. pulling weeds, living with a few bugs around the house, or using natural predators to control pests and organic methods to control weeds). Take an integrated pest management (IPM) approach to controlling undesirable organisms, and you often won't have to use toxic chemicals at all. If you do need to use these chemicals, use only as much as you need; target their application, and don't apply them right before rainfall is expected.
4. Replace your lawn and high-maintenance plants with native plants. Lawns require a lot of water and, generally, a lot of chemicals. The

same can be said for many other plants that aren't necessary suited for survival in your yard. By replacing these high-maintenance plants with native species, you can reduce or eliminate your use of pesticides, herbicides, and fertilizers, and you won't have to spend as much time tending your yard. You can also dramatically lower your water use and help prevent runoff and erosion.

5. Properly store and dispose of chemicals. Many household chemicals and automotive products are extremely toxic both to humans and to other organisms. Protect water quality by making sure these chemicals are stored in tightly sealed containers and that they aren't exposed to extreme temperatures. Clean up spills carefully, rather than leaving them on the ground or washing them into the street. When it comes time to get rid of used or unwanted chemicals, take them to your local hazardous waste recycling facility.
6. Clean up pet waste. Pet waste contains harmful bacteria and other pollutants. While a good rain storm may wash your dog or cat's poop away, it isn't really *gone*—it's in the water supply. Promptly pick up after your pet, and seal the waste in a plastic bag before throwing it in the trash.
7. Contain and/or compost yard waste. Yard waste that sits around can easily wash into storm drains when it rains. Even if the waste doesn't contain chemicals such as herbicides and pesticides, the introduction of large quantities of sticks, leaves, and grass clippings can overwhelm waterways with unhealthy quantities of nutrients. Remember, even beneficial and necessary substances can be harmful if there's too much of them, and waterways can't handle the sudden inflow of mass quantities of organic matter washed down storm drains.
   - Compost yard wastes. Your compost should be contained in a bin or barrel—some municipalities provide these for free or at low cost—to prevent the materials from being washed away.
   - Use a mulching mower instead of bagging grass clippings. Mulching mowers add a natural layer of compost to your lawn, and you don't have to deal with disposal of grass clippings.
   - Dispose of yard and grass clippings properly. If you don't compost or have yard wastes that you can't compost, contact your local waste management or environmental protection agency to determine how to dispose of yard wastes. Many jurisdictions provide regularly schedule yard waste pickups, and others allow you to schedule separate pickups. In any case, bag or otherwise contain the material while you're waiting for pickup.

- o Contain disturbed soil. If your revamping your landscape or tearing out old sod, you can end up with big piles of dirt and organic matter. These are highly susceptible to being washed away in runoff and should therefore be covered or otherwise contained, even if they will only be there for a short time.

8. Pick up litter and properly dispose of trash. Litter isn't just unsightly; it can also contribute to water pollution. Just about every material—from paper to cigarette butts to aluminum cans and old appliances—contains chemicals that can leach out into the environment. Everybody knows that littering is a no-no, but it's important to understand that trash or junk sitting in your yard can be just as harmful as trash illegally dumped by the side of the road.
9. Avoid using salt to de-ice walkways. In colder climates, salting walkways and driveways is a common practice. It's so common, in fact, that freshwater streams and lakes in these areas have been found to have extraordinarily high concentrations of salt—high enough to kill off fish and other aquatic organisms. Regularly and thoroughly shovel and/or sweep snow from your walkways instead of relying on salt, and sparingly apply non-toxic alternatives to salt to surfaces that need de-icing or extra traction. Examples of alternatives to salt include gravel and biodegradable, low-toxicity chemicals such as calcium magnesium acetate and liquid potassium acetate.
10. Maintain your septic system. If you have a septic system, have it regularly inspected and maintained. Overloaded or improperly functioning septic systems can spew raw sewage directly into bodies of water or can contaminate groundwater. Most septic systems should be pumped every 2-3 years.
11. Maintain a vegetated buffer between your yard and bodies of water. If you live near a body of water, keep or plant a buffer of vegetation to capture runoff from your yard. Don't mow your lawn all the way up to the shore, and seriously consider replacing a lawn buffer with native plants. This area should be completely free of pet waste, pesticides, herbicides, or fertilizers. People who live in close proximity to streams, lakes, and oceans have a special responsibility in the fight against water pollution, because they can more directly contaminate these bodies of waters than others who live further away.

Tips

- Hazardous waste isn't limited to chemicals like drain cleaner or gasoline.
- Household products such as electronics, batteries, and thermometers also often contain toxic substances. If you're not sure whether something

is hazardous, check with your local waste management or environmental protection department or do some research online.

- Think about the big picture. You may think that a little oil leak on your car isn't a big deal, and in a way, you're right. The oil from thousands or millions of cars with minor oil leaks, however, adds up quickly, and pretty soon you're looking at a cumulative oil spill far worse than any oil tanker crash. You can't fix all the oil leaks in the world, but you can fix yours. Be part of the solution.
- Educate your family, friends, and neighbors about ways to reduce their contributions to pollution. If your community doesn't already have environmental education programs, pollution control regulations, or a hazardous waste recycling facility, take the initiative to get the ball rolling.
- In many areas, agricultural runoff is a bigger pollution problem than urban runoff. If you're involved in agriculture, contact your local extension service or environmental protection agency to find out more about ways you can reduce your environmental impact.

ENVIRONMENTAL MANGEMENT

The management of interaction of modern human societies with an impact upon the environment.

The three main issues that affect managers are those involving politics (networking), programs (projects) and resources (money, facilities, etc.). The need for environmental management can be viewed from a variety of perspectives.

A more common philosophy and impetus behind environmental management is the concept of carrying capacity. Simply put, carrying capacity refers to the maximum number of organisms a particular resource can sustain.

The concept of carrying capacity, whilst understood by many cultures over history, has its roots in Malthusian theory. Environmental management is therefore not the conservation of the environment solely for the environment's sake, but rather the conservation of the environment for humankind's sake.[citation needed] This element of sustainable exploitation, getting the most out of natural assets, is visible in the EU Water Framework Directive.

Environmental management involves the management of all components of the bio-physical environment, both living (biotic) and non-living (abiotic). This is due to the interconnected and network of relationships amongst all living species and their habitats.

The environment also involves the relationships of the human environment, such as the social, cultural and economic environment with the bio-physical environment. As with all management functions, effective management tools, standards and systems are required. An 'environmental management standard or system or protocol attempts to reduce environmental impact as measured by some objective criteria.

The ISO 14001 standard is the most widely used standard for environmental risk management and is closely aligned to the European Eco-Management and Audit Scheme (EMAS). As a common auditing standard, the ISO 19011 standard explains how to combine this with quality management.

Other environmental management systems (EMS) tend to be based on the ISO 14001 standard and many extend it in various ways:

The Green Dragon Environmental Management Standard is a five level EMS designed for smaller organisations for whom ISO 14001 may be too onerous and for larger organisations who wish to implement ISO 14001 in a more manageable step-by-step approach

BS 8555 is a phased standard that can help smaller companies move to ISO 14001 in six manageable steps

The Natural Step focuses on basic sustainability criteria and helps focus engineering on reducing use of materials or energy use that is unsustainable in the long term

Natural Capitalism advises using accounting reform and a general bio mimicry and industrial ecology approach to do the same thing

US Environmental Protection Agency has many further terms and standards that it defines as appropriate to large-scale EMS.[citation needed]

The UN and World Bank has encouraged adopting a "natural capital" measurement and management framework. The European Union Eco-Management and Audit Scheme (EMAS)

Some personal views and suggestions to encourage a more positive role of the third world in the legislation and implementation of international environmental law:

Firstly, it should be pointed out that, in spite of their low economic and technological levels, the third-world nations have not taken a passive attitude nor have they been indifferent toward international environmental protection. On the contrary, many of them exerted their utmost efforts by taking an active part in international environmental legislation and implementation.

Especially since the 1980s, they have made active appeals for the strengthening of international cooperation; the provision of assistance to expand coastal jurisdiction; protection against sea-coast pollution; the control of the export and transport of hazardous products; the control of the processing of solid wastes; and the protection of the soil. Owing to their active initiative, participation, and support, such instruments as the World Soil Charter, the

World Charter for Nature, and the United Nations Convention on the Law of the Sea were enacted.

Here, it is appropriate to mention that the third-world countries united to ensure the adoption of the United Nations Declaration on the Right to Development (1986), whereby the bases and contents of international environmental laws were further expanded.

Secondly, there is no doubt that what the third-world nations urgently need is to develop their national economies, establish their own modernized industries, improve their agricultural methods, obtain their economic independence, as well as to maintain and consolidate their political independence and sovereignty.

Then and only then will third-world countries be able to exert their utmost efforts towards improving the environment. Such improvement must come gradually, according to their financial and social capacities.

The environmental plight in third-world nations cannot be mentioned in the same breath with that of developed countries. There is a great difference in the damage caused by the latter countries. Even though environmental pollution in some developing states is very serious and affects other states, as well as the world's ecosystem, the attitudes toward such pollution are different.

In the United Nations Declaration on the Human Environment of 1972 it was said that "*... in developing countries most of the environmental problems are caused by underdevelopment... developing countries should devote themselves to development.*"

Experience also shows that only on the basis of self-reliance, after their national economies have been developed, modern industries have been established, and agricultural methods have been improved, can third-world nations gradually achieve effectiveness in maintaining and improving their own environments.

Then and only then can they effectively throw themselves into the struggle for international environmental protection. Needless to say, while engaging in domestic production and construction, third-world nations should interact appropriately with the environment and constantly balancing parochial and global interests, need to take the necessary measures to protect the environment and to control pollution. Thirdly, it should also be mentioned that developed countries should continue to keep a high profile on pollution control, especially since they are the main contributors of environmental pollution. The earth is a single entity and global environmental protection is a matter of life and death for all of humanity.

As a result, all of its members are jointly liable. Developed countries must utilize their economic and technological advantages to create an international environment favourable to economic development and to improve that which the third-world states are unable to do because of their poverty. Developed countries should also do more to help third-world nations financially and technologically, in addition to encouraging them to become involved in efforts to control the global environment.

Lastly, considering the economic and political features of third world states as well as the cause and nature of their environmental pollution, the incentives adopted to attract them to active participation in the drafting and implementing of international environmental legislation should be based on the principles that all nations, big or small, are equally important and that full consultation in any international cooperation or action is imperative. No country is allowed to encroach on another's sovereignty, interfere with another's internal affairs, or impair another's interests under the pretext of environmental protection. Coercion, sanctions, and conditional investments or technical transfers are all unworkable.

TRANSPORT AND CHEMICAL REACTIONS OF WATER POLLUTANTS

Most water pollutants are eventually carried by the rivers into the oceans. In some areas of the world the influence can be traced hundred miles from the mouth by studies using hydrology transport models. Advanced computer models such as SWMM or the DSSAM Model have been used in many locations worldwide to examine the fate of pollutants in aquatic systems. Indicator filter feeding species such as copepods have also been used to study pollutant fates in the New York Bight, for example. The highest toxin loads are not directly at the mouth of the Hudson River, but 100 kilometers south, since several days are required for incorporation into planktonic tissue. The Hudson discharge flows south along the coast due to coriolis force. Further south then are areas of oxygen depletion, caused by chemicals using up oxygen and by algae blooms, caused by excess nutrients from algal cell death and decomposition. Fish and shellfish kills have been reported, because toxins climb the foodchain after small fish consume copepods, then large fish eat smaller fish, etc. Each step up the food chain concentrates certain toxins like heavy metals and DDT by approximately a factor of ten.

For several years ocean researcher Charles Moore has been investigating a concentration of floating plastic debris in the Pacific Ocean. His study indicates that ocean currents have added to the mass until it is now about the size of Texas. Many of these long-lasting pieces wind up in the stomachs of marine birds and animals.

Many chemicals undergo reactive decay or change especially over long periods of time in groundwater reservoirs. A noteworthy class of such chemicals are the chlorinated hydrocarbons such as trichloroethylene (used in industrial metal degreasing) and tetrachloroethylene used in the dry cleaning industry. Both of these chemicals, which are carcinogens themselves, undergo partial decomposition reactions leading to new hazardous chemicals.

Groundwater pollution is much more difficult to abate than surface pollution because groundwater can move great distances through unseen aquifers. Non-porous aquifers such as clays partially purify water of bacteria by simple filtration (adsorption and absorption), dilution, and, in some cases, chemical reactions and biological activity: however, in some cases, the pollutants merely transform to soil contaminants. Groundwater that moves through cracks and caverns is not filtered and can be transported as easily as surface water. In fact this can be aggravated by the human tendency to use natural sinkholes as dumps in areas of Karst topography.

Regulatory Framework

In the UK there are common law rights to protect the passage of water across land unfettered in either quality of quantity. Criminal laws dating back to the 16th century excercised some control over water pollution but it was not until the River (Prevention of pollution) Acts 1951-1961 were enacted that any systematic control over water pollution was established. These laws were strengthened and extended in the Control of Pollution Act 1984 which has since been updated and modified by a series of further acts.

In the USA, concern over water pollution resulted in the enactment of state anti-pollution laws in the latter half of the 19th century, and federal legislation enacted in 1899. The Refuse Act of the federal Rivers and Harbors Act of 1899 prohibits the disposal of any refuse matter from into either the nation's navigable rivers, lakes, streams, and other navigable bodies of water, or any tributary to such waters, unless one has first obtained a permit. The Water Pollution Control Act, passed in 1948, gave authority to the Surgeon General to reduce water pollution.

Growing public awareness and concern for controlling water pollution led to enactment of the Federal Water Pollution Control Act Amendments of 1972. As amended in 1977, this law became commonly known as the Clean Water Act. The Act established the basic mechanisms for regulating contaminant discharge. It established the authority for EPA to implement wastewater standards for industry. The Clean Water Act also continued requirements to set water quality standards for all contaminants in surface waters. Further amplification of the Act continued including the enactment of the Great Lakes Legacy Act of 2002 (Public Law 107-303, November 27, 2002).

4

Environmental Microbial Ecology

Environmental microbiology is the study of the composition and physiology of microbial communities in the environment. The environment in this case means the soil, water, air and sediments covering the planet and can also include the animals and plants that inhabit these areas. Environmental microbiology also includes the study of microorganisms that exist in artificial environments such as bioreactors.

Microbial life is amazingly diverse and microorganisms literally cover the planet.

An average gram of soil contains approximately one billion (1,000,000,000) microbes representing probably several thousand species. Microorganisms have special impact on the whole biosphere. They are the backbone of ecosystems of the zones where light cannot approach. In such zones, chemosynthetic bacteria are present which provide energy and carbon to the other organisms there.

Some microbes are decomposers which have ability to recycle the nutrients. Microbes have a special role in biogeochemical cycles. Microbes, especially bacteria, are of great importance because their symbiotic relationship (either positive or negative) have special effects on the ecosystem.

Microorganisms are used for *in-situ* microbial biodegradation or bioremediation of domestic, agricultural and industrial wastes and subsurface pollution in soils, sediments and marine environments. The ability of each microorganism to degrade toxic waste depends on the nature of each contaminant. Since most sites typically have multiple pollutant types, the most effective approach to microbial biodegradation is to use a mixture of bacterial species and strains, each specific to the biodegradation of one or more types of contaminants. It is vital to monitor the composition of the indigenous and added bacteria in order to evaluate the activity level and to permit modifications of the nutrients and other conditions for optimizing the bioremediation process.

MOLECULAR BIOLOGY APPROACH TO ENVIRONMENTAL MICROBIOLOGY

Just as the advent of magnifying lenses opened a new area in scientific knowledge and research into the hitherto unknown world of microorganisms, advances in molecular biology and genetic studies are now available to the environmental microbiologist. Our research has employed the use of microbiological culture media and biochemical tests in the identification of microbial populations in environmental samples. Also, different types of microscopic methods have been used in our studies. The trend now is on molecular approaches based on nucleic acid sequence analysis for direct measure of the abundance, diversity and phylogeny of microorganisms in the environment. This hereditary material (DNA or 16sRNA) is specific to species that can now be identified in a mixed population that is typical of environmental samples. Also the use of nucleic acid probes allows the determination of the taxonomic status of individual cells in complex assemblages in the ecosystem (Hurst, 1997). This method has the advantage that cells can be identified irrespective of their state of health. Thus, viable and nonculturable bacteria in the population can be included in the ecosystem. Moreover, there will then be little use of pure cultures that do not occur in most environmental samples.

Thus during my visit to Rutgers University, New Brunswick, USA in 2000, I worked in collaboration with Douglas Eveleigh using molecular methods to estimate the microbial biomass of an enclosed environment. Samples were taken from the alimentary tract of the giant African land snail, *Archachatina marginata* Swainson. Bacterial DNA in the samples was extracted for sequencing. The ability of gut bacteria to produce enzymes that breakdown components of snail feedstuff was also investigated. This was studied with a view to elucidating the roles of bacteria in digesting plant materials in snail gut ecosystem. Previous studies at the University of Lagos (Odiete & Akpata, 1983) revealed the presence of two undescribed or 'new' bacteria species namely; *Corynebacterium sp nov.* and *Flavobacterium sp nov.* The results are encouraging in the search of unusual environments for microbes with novel capabilities to meet the demands of industries.

ROLES OF AQUATIC MICROORGANISMS IN ENVIRONMENTAL MANAGEMENT

From the discussion it is clear that aquatic microbes impact on man and the environment in various ways and they play important roles in environmental management. Microbes are used for assessing the quality of water. For instance, if the count of heterotrophic bacteria is high it is a sign of organic matter in a water source, hence the water will be unsuitable for culinary purposes. However there are many uses of water and the quality required depends on the use intended. Microbes are useful in prediction of impacts.

The presence of *E. coli* indicates recent contamination of water with sewage or faeces with a high probability of the presence of pathogens in the water. Therefore it can be predicted that consumers of the water may suffer infection of the gastrointestinal tract.

Isolation of *Clostridium perfringes* or enterococci indicates that the water had been contaminated at sometime. It is suggested that such water be studied to locate the source of pollution and necessary measures for abatement be taken immediately.

In the abatement of pollution, standards are set of microbes that are monitored. Industries and organizations keep within the allowable limits or pay the penalty for non compliance.

Treatment of industrial waste waters or domestic sewage is mandatory. A microcosm of microbes brings about the biodegradation of waste matter in the sewage treatment plant.

Unlike raw sewage, treated effluent is more amenable to the receiving environment on disposal. This shows that microbes are used in setting standards and in sewage treatment. Furthermore, the peculiar biodegradative abilities of some microbes have been employed in the rehabilitation of polluted environments. Some *Pseudomonas* species have been exploited in biological remediation of oil polluted environments because the bacteria can breakdown the crude oil.

Conclusion

Attention has been drawn to the fact that evolution of life started in the aquatic environment and microbes were the first formed living cells. To date some microbes remain in their original habitat of water which is abundant on the surface as well as in the aquifer on planet Earth.

Microbes isolated from water and especially those in unusual or extreme habitats may have potentials that are novel and useful to man.

Water is essential for sustainability of life. Unfortunately, the degradation of water quality is largely by human activities that impede the natural ability of autochthonous microbes to effect self-purification of water. Microorganisms are useful in sewage treatment and in water quality assessment. Knowledge of aquatic microbial ecology may be used in the control of water pollution.

Environmental Impact Assessment (EIA) and maintenance of a healthy aquatic ecosystem would ensure sustainable development.

RECOMMENDATIONS

The responsibilities of a university professor may be grouped into three: Training of high level man power, carrying out research to generate new ideas and knowledge for sustainable development, and participation in the management of academic resources in terms of administration at different levels of the community.

It is customary at the end of an inaugural lecture to send feelers to the public on the relevance of the ivory tower research to national issues. In this lecture some of our research studies were discussed. Under recommendations, I would like to highlight the limitations we face presently in the area of manpower development and also dwell on the need for community involvement in environmental management.

1. Manpower development involves human capacity building. The urriculum for Microbiology at the University of Lagos contains the essential elements required to meet the demands of environmental microbiologists anywhere. However, our students need to be familiar with some newer techniques. They will benefit more from the molecular approach so that they can be more marketable. Unfortunately, it is becoming increasingly difficult to organize enough laboratory practicals to support theory. Even the traditional methods lack space and consumable materials! The University of Lagos was not always like this!! There are too many students for the available spaces, and funds are inadequate for routine laboratory work. The recommendation is that more spaces be provided for practical work so as to enhance our understanding and practice of Microbiology.
2. Environmental Education. The elevation of the Federal Environmental Protection Agency to Federal Ministry of Environment is in line with global trends. Nigeria is playing a leadership role in environmental matters in Africa. Indeed the country is represented at international meetings on the Environment. However, this top administrative posture must be matched with grass-root knowledge and participation in environmental management. Individuals should recognize early signs of environmental degradation and know those to whom a report would receive quick response. Environmental issues are largely what are seen rather than what is said or written! In other words, emphasis should be on implementation of policies. Our leadership status should be reflected in the level of cleanliness and health of the ecosystem. To achieve this, Environmental Education must be implemented in the curricula of primary and secondary schools. This would infuse environmental awareness into our value system such that environmental

consciousness becomes a national culture and a clean environment the pride of all Nigerians.

MICROBIAL EOLOGY

Microbial ecology is the relationship of microorganisms with one another and with their environment. It concerns the three major domains of life — Eukaryota, Archaea, and Bacteria — as well as viruses. Microorganisms, by their omnipresence, impact the entire biosphere. They are present in virtually all of our planet's environments, including some of the most extreme, from acidic lakes to the deepest ocean, and from frozen environments to hydrothermal vents.

Microbes, especially bacteria, often engage in symbiotic relationships (either positive or negative) with other organisms, and these relationships affect the ecosystem. One example of these fundamental symbioses are chloroplasts, which allow eukaryotes to conduct photosynthesis. Chloroplasts are considered to be endosymbiotic cyanobacteria, a group of bacteria that are thought to be the origins of aerobic photosynthesis. Some theories state that this invention coincides with a major shift in the early earth's atmosphere, from a reducing atmosphere to an oxygen-rich atmosphere. Some theories go as far as saying that this shift in the balance of gasses might have triggered a global ice-age known as the Snowball Earth. They are the backbone of all ecosystems, but even more so in the zones where light cannot approach and thus photosynthesis cannot be the basic means to collect energy. In such zones, chemosynthetic microbes provide energy and carbon to the other organisms. Other microbes are decomposers, with the ability to recycle nutrients from other organisms' waste poducts. These microbes play a vital role in biogeochemical cycles.

The nitrogen cycle, the phosphorus cycle and the carbon cycle all depend on microorganisms in one way or another. For example, nitrogen which makes up 78% of the planet's atmosphere is "indigestible" for most organisms, and the flow of nitrogen into the biosphere depends on a microbial process called fixation.

Due to the high level of horizontal gene transfer among microbial communities, microbial ecology is also of importance to studies of evolution.

MICROBIAL RESOURCE MANAGEMENT

Biotechnology may be used alongside microbial ecology to address a number of environmental and economic challenges. Managing the carbon cycle to sequester carbon dioxide and prevent excess methanogenesis is important in mitigating global warming, and the prospects of bioenergy are being expanded by the development of microbial fuel cells. Microbial resource management advocates a more progressive attitude towards disease, whereby biological control agents are favoured over attempts at eradication. Fluxes in microbial communities has to be better characterized for this field's potential to be realised.

ROLE OF MICRO-ORGANISMS IN BIOGEOCHEMICAL CYCLES

Biogeochemical Cycles

While humans cannot control the weather on a daily basis, the influence of human life on the environment plays a significant role in global climate. How often have you wished for a rainy day to go away, or for the warm weather of summer during wintertime? Unfortunately, these wishes rarely come true, and it seems as though humans have little control over the weather. While humans cannot control the weather on a daily basis, the influence of human life on the environment plays a significant role in global climate. Deforestation and fossil fuel burning are just a couple of examples of human activities that seriously disrupt the equilibrium of the global ecosystem and alter the biogeochemical cycles that play a role in determining the Earth's climate.

Biogeochemical cycles are essentially the continuous transport and transformation of materials in the environment. Materials are transported through life, air, sea and land in a series of cycles. These cycles include the circulation of elements and nutrients upon which life and the earth's climate depend.

The most important biogeochemical cycles are those of water, carbon, nitrogen and certain other trace gases. In this text, however, we will discuss the carbon and nitrogen cycles, as they are closely intertwined with living things on Earth.

The carbon cycle is particularly influential when it comes to global climate. Much of the carbon in the carbon cycle is in the form of carbon dioxide, a gas that has a strong greenhouse effect because it absorbs infrared radiation.

Carbon is one of the most common elements on Earth and it is the basis of all living things. Below is a graphical depiction of the carbon cycle:

These carbon compounds either:

a) decay into peat, then over millions of years, coal (under very high pressures and worked on by microbes in the absence of oxygen). The coal is then burned by factories to produce electricity, and thus the carbon is returned to carbon dioxide in the air, or

b) are eaten by animals (or remain in the plant, no difference). The carbon compounds in both the plants and animals are returned to the air as carbon dioxide via respiration and also when they die and decay, as microbes digest their biomass.

Nitrogen is another element that plays important roles in both biological and non-biological systems. Nitrogen gas makes up 80% of the Earth's atmosphere and nitrogen exists in proteins of living organisms. Nitrogen in the air is built up into nitrates by nitrogen fixing bacteria. These nitrates are then absorbed by plants and turned into plant proteins. Leguminous plants can simply take the nitrogen in the air, and then build it up into plant proteins. The plant protein is then eaten by animals, who then excrete the protein as ammonia. Both the plant and animals proteins can be broken down and digested by microbes once the plant or animal dies into ammonia.

This ammonia is then oxidized by nitrifying bacteria into nitrites, which are then oxidized again by other nitrifying bacteria into nitrates.

Denitrifying bacteria can reduce nitrates to nitrogen in the air, nitrites or ammonia. Global climate change, temperature, precipitation and the stability of ecosystems are all dependent upon biogeochemical cycles. When humans inadvertently disrupt these cycles by, for example, polluting, disastrous consequences can result. A healthy understanding of these cycles are critical in order to ensure the health and safety of future generations of living things on Earth. From climate changes to atmospheric composition, biogeochemical cycles are an integral component of planetary biology.

BIODEGRADATION OF POLLUTANTS

Microbial biodegradation of pollutants plays a pivotal role in the bioremediation of contaminated soil and ground-water sites. Such pollutants include chloroethenes, steroids, organophosphorus compounds, alkanes, PAHs and PCBs.

Oil Biodegradation

Petroleum oil is toxic, and pollution of the environment by oil causes major ecological concern. Oil spills of coastal regions and the open sea are poorly containable and mitigation is difficult; much of the oil can, however, be eliminated by the hydrocarbon-degrading activities of microbial communities, in particular the hydrocarbonoclastic bacteria (HCB). These organisms can help remedy the ecological damage caused by oil pollution of marine habitats. HCB also have potential biotechnological applications in the areas of bioplastics and biocatalysis.

Waste Biotreatment

Biotreatment, the processing of wastes using living organisms, is an environmentally friendly alternative to other options for treating waste material.

Bioreactors have been designed to overcome the various limiting factors of biotreatment processes in highly controlled systems. This versatility in the design of bioreactors allows the treatment of a wide range of wastes under optimized conditions. It is vital to consider various microorganisms and a great number of analyses are often required.

Wastewater Treatment

Wastewater treatment processes are geared towards one purpose: cleaning up water. Recent application of molecular techniques is unveiling the microbial composition and architecture of the complex communities involved in the treatment processes. It is now recognized that wastewater processes harbour a vast variety of microorganisms most of which are yet-to-be cultured, hence uncharacterized. Metagenomic technology is being used to study the diversity, structure and functions of microbial communities in nitrifying processes, anaerobic ammonia oxidation processes and methane fermenting processes.

Environmental Genomics of Cyanobacteria

The application of molecular biology and genomics to environmental microbiology has led to the discovery of a huge complexity in natural communities of microbes. Diversity surveying, community fingerprinting and functional interrogation of natural populations have become common, enabled by a range of molecular and bioinformatics techniques.

Recent studies on the ecology of Cyanobacteria have covered many habitats and have demonstrated that cyanobacterial communities tend to be habitat-specific and that much genetic diversity is concealed among morphologically simple types. Molecular, bioinformatics, physiological and geochemical techniques have combined in the study of natural communities of these bacteria.

Corynebacteria

Corynebacteria are a diverse group Gram-positive bacteria found in a range of different ecological niches such as soil, vegetables, sewage, skin, and cheese smear. Some, such as *Corynebacterium diphtheriae*, are important pathogens while others, such as *Corynebacterium glutamicum*, are of immense industrial importance. *C. glutamicum* is one of the biotechnologically most important bacterial species with an annual production of more than two million tons of amino acids, mainly L-glutamate and L-lysine.

Legionella

Legionella is common in many environments, with at least 50 species and 70 serogroups identified. *Legionella* is commonly found in aquatic habitats where its ability to survive and to multiply within different protozoa equips the bacterium to be transmissible and pathogenic to humans.

Archaea

Originally, Archaea were once thought of as extremophiles existing only in hostile environments but have since been found in all habitats and may contribute up to 20% of total biomass. Archaea are particularly common in the oceans, and the archaea in plankton may be one of the most abundant groups of organisms on the planet. Archaea are subdivided into four phyla of which two, the Crenarchaeota and the Euryarchaeota, are most intensively studied.

Lactobacillus

Lactobacillus species are found in the environment mainly associated with plant material. They are also found in the gastrointestinal tract of humans, where they are symbiotic and make up a portion of the gut flora.

Aspergillus

Aspergillus spores are common components of aerosols where they drift on air currents, dispersing themselves both short and long distances depending on environmental conditions. When the spores come in contact with a solid or liquid surface, they are deposited and if conditions of moisture are right, they germinate. The ability to disperse globally in air currents and to grow almost anywhere when appropriate food and water are available means that ubiquitous is among the most common adjectives used to describe these moulds.

Microbial Nitrogen Cycling

Microorganisms that convert gaseous nitrogen (N_2) to a form suitable for use by living organisms are pivotal for life on earth. This process is called nitrogen fixation. Another set of microbial reactions utilise the bioavailable nitrogen creating N_2 and completing the cycle in a process called denitrification. This crucial nutrient cycle has long been the subject of extensive research.

Rhizobia

Symbiotic nitrogen fixation is a mutualistic process in which bacteria reside inside plants and reduce atmospheric nitrogen to ammonia. This ammonia can then be used by the plant for the synthesis of proteins and other nitrogen-containing compounds such as nucleic acids. The Gram-negative soil bacteria that carry out this process are collectively referred to as rhizobia (from the Greek words Riza = Root and Bios = Life). The process of symbiotic nitrogen fixation is of agricultural and ecological significance because plants capable of nitrogen fixation do not need to compete for limited quantities of soil nitrogen, nor do they require expensive nitrogenous fertilizers that can be harmful to the environment.

Microalgae

Algae are a highly diverse group of protists, ranging from simple, unicellular organisms to complex, multicellular entities with a range of differentiated tissues and distinct organs. They are found among diverse aquatic ecosystems and play important roles by supplying carbon and energy as well as providing habitat to other members of the biological communities. Some algae cause significant environmental and health problems. There are three algal groups: the dinoflagellates, the diatoms and the haptophytes. Their 3 main phylla are chlorophyta,rhodophyta and phaeophyta.

Anaerobic Protozoa

Diplomonads are a group of mitochondrion-lacking, binucleated flagellates found in anaerobic or micro-aerophilic environments. Most research on diplomonads has focused on *Giardia,* which is a major cause of water-borne enteric disease in humans and other animals. The first diplomonad to have its genome sequenced was a *Giardia* isolate (WB) and the 11.7 million basepair genome is compact in structure and content with simplified basic cellular machineries and metabolism.

Water Microbiology

An adequate supply of safe drinking water is one of the major prerequisites for a healthy life, but waterborne diseases are still a major cause of death in many parts of the world, particularly in young children, the elderly, or those with compromised immune systems. As the epidemiology of waterborne diseases is changing, there is a growing global public health concern about new and reemerging infectious diseases that are occurring through a complex interaction of social, economic, evolutionary, and ecological factors. An important challenge is therefore the rapid, specific and sensitive detection of waterborne pathogens. Presently, microbial tests are based essentially on time-consuming culture methods. However, newer enzymatic, immunological and genetic methods are being developed to replace and/or support classical approaches to microbial detection. Moreover, innovations in nanotechnology and nanosciences are having a significant impact in biodiagnostics, where a number of nanoparticle-based assays and nanodevices have been introduced for biomolecular detection.

Molecular techniques based on genomics, proteomics and transcriptomics are rapidly growing as complete microbial genome sequences are becoming available, and advances are made in sequencing technology, analytical biochemistry, microfluidics and data analysis. While the clinical and food industries are increasingly adapting these techniques, there appear to be major challenges in detecting health-related microbes in source and treated drinking waters. This is due in part to the low density of pathogens in water, necessitating significant processing of large volume samples. From the vast panorama of available molecular techniques, some are finding a place in the water industry: Quantitative PCR, protein detection and immunological approaches, loop-mediated isothermal amplification (LAMP), microarrays.

Sensory Mechanisms

Bacteria have evolved abilities to regulate aspects of their behaviour (such as gene expression) in response to signals in the intracellular and extracellular environment. The interaction of a signal with its receptor (usually a protein or RNA molecule) triggers a series of events that lead to reprogramming of cellular physiology, typically as a consequence of altered patterns of gene expression. In this way, the bacterial cell is able to mount appropriate and effective responses to changing physical and/or chemical environments.

The versatility with which many bacteria adapt to environmental change underlies many important aspects of microbiology. For example, pathogens encounter multiple environments as they invade a host from the outside, and then progress through different sites within host tissues. There is growing evidence that pathogenic bacteria make use of physical and chemical cues to signal their presence in a suitable host, and need to adapt to the host environment in order to mount a successful infection. On the other hand, it should not be assumed that all signals to which bacteria must respond originate in the extracellular environment. For many species, even the cosseted life in a laboratory shake flask is 'stressful', in the sense that there is often a need to avoid or reverse the effects of harmful intermediates or by-products of metabolism. For example, all organisms that use dioxygen as a terminal electron acceptor have to deal with the reactive oxygen species that arise as adventitious by-products of aerobic metabolism. In bacteria, multiple protein receptors for oxygen radicals have been described, which control the expression of genes encoding enzymes that detoxify oxygen radicals or repair the damage that they cause.

Metagenomics

Metagenomics is the cultivation-independent analysis of the collective genomes of microbes within a given environment, using sequence- and function-based approaches. Metagenomic studies have revealed the vast size and richness of the microbial and viral world and demonstrated the phylogenetic diversity of various environments.

Access to huge volume genomic sequence data from uncultured organisms has opened up many new avenues of research. Advances in the throughput of sequencing and screening technologies have greatly facillitated metagenomics research.

EMERGING TECHNOLOGIES: BIOREPORTERS, BIOSENSORS AND MICROPROBES

Bioreporter

Bioreporters are intact, living microbial cells that have been genetically engineered to produce a measurable signal in response to a specific chemical or physical agent in their environment. Bioreporters contain two essential genetic elements, a promoter gene and a reporter gene. The promoter gene is turned on (transcribed) when the target agent is present in the cell's environment. The promoter gene in a normal bacterial cell is linked to other genes that are then likewise transcribed and then translated into proteins that help the cell in either combating or adapting to the agent to which it has been exposed. In the case of a bioreporter, these genes, or portions thereof, have been removed and replaced with a reporter gene. Consequently, turning on the promoter gene now causes the reporter gene to be turned on. Activation of the reporter gene leads to production of reporter proteins that ultimately generate some type of a detectable signal. Therefore, the presence of a signal indicates that the bioreporter has sensed a particular target agent in its environment.

Originally developed for fundamental analysis of factors affecting gene expression, bioreporters were early on applied for the detection of environmental contaminants and have since evolved into fields as diverse as medical diagnostics, precision agriculture, food safety assurance, process monitoring and control, and bio-microelectronic computing. Their versatility stems from the fact that there exist a large number of reporter gene systems that are capable of generating a variety of signals. Additionally, reporter genes can be genetically inserted into bacterial, yeast, plant, and mammalian cells, thereby providing considerable functionality over a wide range of host vectors.

REPORTER GENE SYSTEMS

Several types of reporter genes are available for use in the construction of bioreporter organisms, and the signals they generate can usually be categorized as either colorimetric, fluorescent, luminescent, chemiluminescent or electrochemical. Although each functions differently, their end product always remains the same – a measurable signal that is proportional to the concentration of the unique chemical or physical agent to which they have been exposed. In some instances, the signal only occurs when a secondary substrate is added to the bioassay (*luxAB*, Luc, and aequorin).

For other bioreporters, the signal must be activated by an external light source (GFP and UMT), and for a select few bioreporters, the signal is completely self-induced, with no exogenous substrate or external activation being required (*luxCDABE*). The following sections outline in brief some of the reporter gene systems available and their existing applications.

Bacterial Luciferase (Lux)

Luciferase is a generic name for an enzyme that catalyzes a light-emitting reaction. Luciferases can be found in bacteria, algae, fungi, jellyfish, insects, shrimp, and squid, and the resulting light that these organisms produce is termed bioluminescence. In bacteria, the genes responsible for the light-emitting reaction (the *lux* genes) have been isolated and used extensively in the construction of bioreporters that emit a blue-green light with a maximum intensity at 490 nm. Three variants of *lux* are available, one that functions at < 30°C, another at < 37°C, and a third at < 45°C. The *lux* genetic system consists of five genes, *luxA*, *luxB*, *luxC*, *luxD*, and *luxE*. Depending on the combination of these genes used, several different types of bioluminescent bioreporters can be constructed.

LuxAB Bioreporters

luxAB bioreporters contain only the *luxA* and *luxB* genes, which together are responsible for generating the light signal. However, to fully complete the light-emitting reaction, a substrate must be supplied to the cell. Typically, this occurs through the addition of the chemical decanal at some point during the bioassay procedure. Numerous *luxAB* bioreporters have been constructed within bacterial, yeast, insect, nematode, plant, and mammalian cell systems.

LuxCDABE Bioreporters

Instead of containing only the *luxA* and *luxB* genes, bioreporters can contain all five genes of the *lux* cassette, thereby allowing for a completely independent light generating system that requires no extraneous additions of substrate nor any excitation by an external light source. So in this bioassay, the bioreporter is simply exposed to a target analyte and a quantitative increase in bioluminescence results, often within less than one hour. Due to their rapidity and ease of use, along with the ability to perform the bioassay repetitively in real-time and on-line, makes *luxCDABE* bioreporters extremely attractive. Consequently, they have been incorporated into a diverse array of detection methodologies ranging from the sensing of environmental contaminants to the real-time monitoring of pathogen infections in living mice.

Nonspecific *lux* Bioreporters

Nonspecific *lux* bioreporters are typically used for the detection of chemical toxins. They are usually designed to continuously bioluminesce. Upon exposure to a chemical toxin, either the cell dies or its metabolic activity is retarded, leading to a decrease in bioluminescent light levels. Their most familiar application is in the Microtox assay where, following a short exposure to several concentrations of the sample, the decreased bioluminescence can be correlated to relative levels of toxicity.

Firefly Luciferase (Luc)

Firefly luciferase catalyzes a reaction that produces visible light in the 550 – 575 nm range. A click-beetle luciferase is also available that produces light at a peak closer to 595 nm. Both luciferases require the addition of an exogenous substrate (luciferin) for the light reaction to occur.

Numerous *luc*-based bioreporters have been constructed for the detection of a wide array of inorganic and organic compounds of environmental concern. Their most promising application, however, probably lies within the field of medical diagnostics. Insertion of the *luc* genes into a human cervical carcinoma cell line (HeLa) illustrated that tumor-cell clearance could be visualized within a living mouse by simply scanning with a charge-coupled device camera, allowing for chemotherapy treatment to rapidly be monitored on-line and in real-time. In another example, the *luc* genes were inserted into human breast cancer cell lines to develop a bioassay for the detection and measurement of substances with potential estrogenic and antiestrogenic activity.

Aequorin

Aequorin is a photoprotein isolated from the bioluminescent jellyfish *Aequorea victoria*. Upon addition of calcium ions (Ca2+) and coelenterazine, a reaction occurs whose end result is the generation of blue light in the 460 - 470 nm range. Aequorin has been incorporated into human B cell lines for the detection of pathogenic bacteria and viruses in what is referred to as the CANARY assay (Cellular Analysis and Notification of Antigen Risks and Yields). The B cells are genetically engineered to produce aequorin. Upon exposure to antigens of different pathogens, the recombinant B cells emit light as a result of activation of an intracellular signaling cascade that releases calcium ions inside the cell.

Green Fluorescent Protein (GFP)

Green fluorescent protein (GFP) is also a photoprotein isolated and cloned from the jellyfish *Aequorea victoria*. Variants have also been isolated from the sea pansy *Renilla reniformis*.

GFP, like aequorin, produces a blue fluorescent signal, but without the required addition of an exogenous substrate. All that is required is an ultraviolet light source to activate the fluorescent properties of the photoprotein.

This ability to autofluoresce makes GFP highly desirable in biosensing assays since it can be used on-line and in real-time to monitor intact, living cells.

Additionally, the ability to alter GFP to produce light emissions besides blue (i.e., cyan, red, and yellow) allows it to be used as a multianalyte detector. Consequently, GFP has been used extensively in bioreporter constructs within bacterial, yeast, nematode, plant, and mammalian hosts.

Uroporphyrinogen (Urogen) III Methyltransferase (UMT)

Uroporphyrinogen (urogen) III methyltransferase (UMT) catalyzes a reaction that yields two fluorescent products which produce a red-orange fluorescence in the 590 - 770 nm range when illuminated with ultraviolet light. So as with GFP, no addition of exogenous substrates is required. UMT has been used as a bioreporter for the selection of recombinant plasmids, as a marker for gene transcription in bacterial, yeast, and mammalian cells, and for the detection of toxic salts such as arsenite and antimonite.

BIOSENSOR

A biosensor is an analytical device for the detection of an analyte that combines a biological component with a physicochemical detector component.

It consists of 3 parts:

- the *sensitive biological element* (biological material (e.g. tissue, microorganisms, organelles, cell receptors, enzymes, antibodies, nucleic acids, etc.), a biologically derived material or biomimic) The sensitive elements can be created by biological engineering.
- the *transducer* or the *detector element* (works in a physicochemical way; optical, piezoelectric, electrochemical, etc.) that transforms the signal resulting from the interaction of the analyte with the biological element into another signal (i.e., transducers) that can be more easily measured and quantified;
- associated electronics or signal processors that are primarily responsible for the display of the results in a user-friendly way.. This sometimes accounts for the most expensive part of the sensor device, however it

is possible to generate a user friendly display that includes transducer and sensitive element.

A common example of a commercial biosensor is the blood glucose biosensor, which uses the enzyme glucose oxidase to break blood glucose down. In doing so it first oxidizes glucose and uses two electrons to reduce the FAD (a component of the enzyme) to FADH2.

This in turn is oxidized by the electrode (accepting two electrons from the electrode) in a number of steps. The resulting current is a measure of the concentration of glucose. In this case, the electrode is the transducer and the enzyme is the biologically active component.

Recently, arrays of many different detector molecules have been applied in so called electronic nose devices, where the pattern of response from the detectors is used to fingerprint a substance.. In the Wasp Hound odor-detector, the mechanical element is a video camera and the biological element is five parasitic wasps who have been conditioned to swarm in response to the presence of a specific chemical. Current commercial electronic noses, however, do not use biological elements.

A canary in a cage, as used by miners to warn of gas, could be considered a biosensor. Many of today's biosensor applications are similar, in that they use organisms which respond to toxic substances at a much lower concentrations than humans can detect to warn of the presence of the toxin. Such devices can be used in environmental monitoring, trace gas detection and in water treatment facilities.

BACTERIAL GENOME EVOLUTIONARY DYNAMICS

Gene Loss / Genome Decay

Gene loss or genome decay occurs when a gene is no longer used by the microbe or when a microbe attempts to adapt to a new ecological niche.

Sequencing efforts and microarray analysis have exposed a large number of pseudo genes in some bacterial pathogen species. Mycobacterium leprae for example has been found to contain nearly as many pseudo genes as function genes. M. leprae is not the only microbe exhibiting such behaviour; in his article, Dr. Pallen reports similar properties from Yersinia pestis (the plague pathogen) and also Salmonella enterica. The inactivation of genes is typically associated with a change in the lifestyle of an organism, which can involve adapting to a new niche. The presence of extensive pseudogenes is contrary to another orthodox belief that all the genes in a bacterial genome are functional for some purpose.

It is possible to detect the presence of pseudogenes and the marks of genome decay through whole-genome sequencing in combination with comparative genomics Comparative genomics has helped to reveal that pathogens may favour losing genes in order to live in a host-associated niche and become endosymbionts. Sometimes the shedding of certain genes also renders a pathogen microbe harmless. The analysis of Listeria strains, for example, has shown that a reduced genome size has led to the generation of a non-pathogenic Listeria strain from a pathogenic precursor.

Gene Gain/ Gene Duplication

One of the key forces driving gene gain is thought to be horizontal (lateral) gene transfer (LGT). It is of particular interest in microbial studies because these mobile genetic elements may introduce virulence factors into a new genome. An important comparative study conducted by Gill et al. in 2005 postulated that LGT may have been the cause for pathogen variations between Staphylococcus epidermidis and Staphylococcus aureus. There still, however, remains scepticism about the frequency of LGT, its identification, and its impact.

New and improved methodologies have been engaged, especially in the study of phylogenetics, to validate the presence and effect of LGT.

Gene gain and gene duplication events are balanced by gene loss, such that despite their dynamic nature, the genome of a bacterial species remains approximately the same size.

Genome Rearrangement

Mobile genetic insertion sequences can play a role in genome rearrangement activities. Pathogens that do not live in an isolated environment have been found to contain a large number of insertion sequence elements and various repetitive segments of DNA. The combination of these two genetic elements is thought help mediate homologous recombination. There are pathogens, such as Burkholderia mallei, and Burkholderia pseudomallei which have been shown to exhibit genome-wide rearrangements due to insertion sequences and repetitive DNA segments. At this time, no studies demonstrate genome-wide rearrangement events directly giving rise to pathogenic behaviour in a microbe. This does not mean it is not possible. Genome-wide rearrangements do, however, contribute to the plasticity of bacterial genome, which may prime the conditions for other factors to introduce, or lose, virulence factors.

Single Nucleotide Polymorphism

Single nucleotide polymorphisms(SNPs) are also a genomic variable that adds to the diversity of pathogen strains. Current efforts attempt to catalogue the various SNPs in pathogen strains.

Analysis of Genomic Diversity

There is a need to analyze more than a single genome sequence of a pathogen species to understand pathogen mechanisms. Comparative genomics is a powerful methodology that has gained more applicability with the recent increased amount of sequence information. There are several examples of successful comparative genomics studies, among them the analysis of Listeria. and Escherichia coli. The most important topic comparative genomics, in a pathogenomic context, attempts to address the difference between pathogenic and non-pathogenic microbes. This inquiry, though, proves to be very difficult to analyze, since a single bacterial species can have many strains and the genomic content of each of these strains can vary.

Pan-genomes and Core Genomes

The diversity within pathogen genomes makes it difficult to identify the total number of genes that are associated within all strains of a pathogen species. In fact, it has been thought that the total number of genes associated with a single pathogen species may be unlimited, although some groups are attempting to derive a more empirical value.. For this reason it was necessary to introduce the concept of pan-genomes and core genomes. Pan-genome and core genome literature also tends to have a bias towards reporting on prokaryotic pathogen organisms. Caution may need to be exercised when extending the definition of a pan-genome or a core-genome to the other pathogen organisms; this is because there is no formal evidence of the properties of these pan-genomes. Here, it will be assumed that the definitions may in fact extend, since all pathogen organisms share in the same dynamic genomic events and rely upon variability within strains as a mechanism of survival and virulence.

A core genome is the set of genes found across all strains of a pathogen species. A pan-genome is the entire gene pool for that pathogen species, and includes genes that are not shared by all strains. Pan-genomes may be open or closed depending on whether comparative analysis of multiple strains reveals no new genes (closed) or many new genes (open) compared to the core genome for that pathogen species. In the open pan-genome, genes may be further characterized as dispensable or strain specific. Dispensable genes are those found in more than one strain, but not in all strains, of a pathogen species. Strain specific genes are those found only in one strain of a pathogen species. The differences in pan-genomes are reflections of the life style of the organism. For example, Streptococcus agalactiae, which exists in diverse biological niches, has a broader pan-genome when compared with the more environmentally isolated Bacillus anthracis. Comparative genomics approaches are also being used to understand more about the pan-genome.

Mobile Genetic Elements that Encode Virulence Factors

Three genetic elements of human-affecting pathogens contribute to the transfer of virulence factors: plasmids, pathogenicity island, and prophages. Pathogencity islands and their detection are the focous of several bioinformatics efforts involved in pathogènomics.

Analyzing Microbe-Microbe Interactions

Microbe-host interactions tend to overshadow the consideration of microbe-microbe interactions. Microbe-microbe interactions though can lead to chronic states of infirmity that are difficult to understand and treat.

Bioflims

Biofilms are an example of microbe-microbe interactions and are thought to be associated with up to 80% of human infections. Recently it has been shown that there are specific genes and cell surface proteins involved in the formation of biofilm.These genes and also surface proteins may be characterized through in silico methods to form an expression profile of biofilm interacting bacteria. This expression profile may be used in subsequent analysis of other microbes to predict biofilm microbe behaviour, or to understand how to dismantle biofilm formation.

AQUATIC MICROBES SHAPE OUR WORLD

The Earth is a complex and ever-changing combination of land, air, water and living organisms. These components are intricately linked so that processes occurring in one affect the others. Microbes play vital roles in this global system by driving biogeochemical cycles that affect the climate and move elements around the planet, making them available to other organisms. Without microbes, the Earth would be uninhabitable and yet they are the least understood life-forms on our planet.

The Marine and Freshwater Microbial Biodiversity programme has investigated how aquatic microbes shape the global system. Marine microbes play a critical role in regulating natural and man-made gases that affect climate. New research has let us cultivate previously-unknown bacteria that break down methybromide - a chemical that destroys stratospheric ozone. Scientists have investigated microbial methane production, and how marine viruses affect levels of dimethyl sulphide, a gas that promotes cloud formation. Other groups have studied how bacteria fix gaseous nitrogen for growth in seawater, and how they interact with trace metals in freshwater sediments, contaminated estuaries and coastal seas. Methyl bromide gas destroys a massive 25% of ozone lost from the stratosphere. It is released when biomass burns, when cars use leaded petrol, and some countries use it to fumigate soils and harvested crops. Researchers at the University of Warwick and the Plymouth Marine Laboratory have cultivated three new types of bacteria with enzymes that break down the gas. Using molecular probes, they discovered that marine environments harbour many different bacteria with this enzyme.

Methane is increasing in the atmosphere faster than carbon dioxide, and has 21 times its global warming potential. Methane is produced by archaea in the absence of oxygen, and researchers usually assume most of it comes from freshwater environments like wetlands and marshes. Scientists from Cardiff University used tracer and stable isotope techniques to study methane production at 21 sites, ranging from full freshwater to seawater. The team's results will help identify methane sources – a critical step towards reducing inputs to the atmosphere.

Marine micro-organisms also produce beneficial gases, for example dimethyl sulphide (DMS), the dominant sulphur trace gas in seawater. In laboratory culture and field mesocosm studies, researchers at the Marine Biological

Association and the University of East Anglia showed that when viruses kill the marine phytoplankton *Emiliania huxleyi*, DMS production increases. DMS emitted from the sea is important in the global sulphur cycle. It also influences global climate by producing sulphate particles that can promote cloud formation, and hence altering how much of the Sun's energy the Earth reflects back to space.

Microbes can convert nitrogen gas to ammonia and organic nitrogen compounds – a critical process in the global nitrogen cycle. Levels of available nitrogen limit phytoplankton growth in the sea, yet more phytoplankton grow in some tropical and sub-tropical regions than known nitrogen inputs can support. Researchers from the Universities of Stirling and Newcastle used molecular methods to show that as-yet-uncultured nitrogen-fixing bacteria are widely distributed. Scientists used to think that microrganisms only fix nitrogen when they have to, because the process uses a lot of energy and other resources such as iron. So it was surprising to discover that key nitrogen-fixation genes in these bacteria were still operating when ammonia and nitrate were available in the seawater.

Living organisms need trace metals for various processes, but elevated levels can be toxic. Researchers from the Plymouth Marine Laboratory and the University of Southampton looked at how copper and zinc affected the bacterial assemblage, and the size and shape of aggregates (decaying phytoplankton debris and bacteria) in water from clean and metal-contaminated estuaries. They also used micro-electrodes to map oxygen concentration in marine aggregates. Scientists from the University of Lancaster developed a 2D imaging technique to assess how microbes influence geochemical processes in sediments. These projects help show how bacterial activity affects the mobility of trace metals at micro scales.

AQUATIC MICROBES: IMPACT ON MAN AND ENVIRONMENT

The aquatic environment is that part of the earth covered with water and it is alled the hydrosphere. Air is the atmosphere while soil is the lithosphere. ydrosphere, atmosphere and lithosphere comprise the biosphere, i.e. an nvironment where living things are found. Microbes are minute living hings that occur in all environments and are therefore ubiquitous. The environment has received more attention globally and locally in the last decade after the United Nations Earth Summit in Rio de Janeiro, 1992. As Professor of Environmental Microbiology, the biosphere is my research domain. But of particular interest to me is the hydrosphere, my specialty being aquatic microbial ecology and water pollution. So, I have chosen for my inaugural lecture the topic, "Aquatic microbes: impact on man and environment".

Microorganisms in the aquatic environment affect man either positively or negatively, directly or indirectly. In other words, these minute living things may be beneficial or harmful to man, his activities as well as other natural resources. For purposes of completeness, it is noted that the presence of some microorganisms in water may sometimes have no apparent or visible effect, being neutral to the ecosystem. The ecosystem is the highest organization in the hierarchy of life. It covers individual organisms in the food chain, populations, energy flow, biogeochemical cycles, interactions between living and non- living resources, as well as human impact. Microorganisms are involved in all of these activities hence man should take cognizance of microorganisms in sustainable development.

Discovery of Microorganisms

For a long time man was not aware of the existence of the microbial world. At that time, activities and impacts of microbes were attributed to the supernatural. The reason is not far fetched. A microbe is invisible to the unaided human eye. In other words our limit of vision does not cover the size of a microbe. We can all visualize one metre length, one centimetre i.e. one hundredth of a metre. But a smaller division of a metre such as one millionth or billionth is not discernable by the unaided human eye. Thanks to Physics and the advent of magnifying lenses. The microscope is used to enlarge cells so that the morphology or shape of a microbial cell can be discerned. It was Antony van Leewenhoeck (1677) who in 1676 made crude microscopes and, out of curiosity, used them to examine various specimens including well- and sea- water. He discovered what he called 'animacules' or 'minute animals'. van Leewenhoeck thus initiated the study of aquatic microbes and is indeed the father of Environmental Microbiology. Whereas a bacterial cell is not visible by the naked human eye, a population of bacteria may become visible after the individual cells are cultured having been supplied with nutrients for growth and multiplication to the size of a mass of cells or a colony. Even today, it is not easy to appreciate and acknowledge the presence of microorganisms. Man tends to relate to the effect, result or symptoms produced by the activities of microorganisms.

What of fermentation of palm juice into palm wine (Odeyemi & Akpata,1988)? Salmonellosis contracted from poultry (Akpata & Fashola, 1985)? Anyone would give prompt attention to gastroenteritis (upset stomach) from cholera or dysentery due to drinking contaminated water or eating molluscs harvested from a polluted bay. If microorganisms cause changes that are pleasant, the process is maintained, if undesirable or harmful, the process is terminated or controlled.

Position of Aquatic Microbes in the Evolution of Life

Let us look at microbes in water, their evolution and niches. Water as a chemical (H2O) is devoid of carbon/energy and nitrogen sources of nutrients required for the growth of living things. It is the presence of extraneous substances in water that provides the nutrients for microbial growth. These microbes affect the quality of water and thereby influence the use that man makes of it.

There are microorganisms in water. Some are indigenous to water and are called autochthonous species while some are foreign or allochthonous species that find their way into the aquatic environment. Autochthonous species are involved in the cleansing of water, utilizing the little nutrients available because a clean body of water is referred to as oligotrophic, having low nutrient content. When wastes are loaded into water it becomes eutrophic providing nutrients for microbial proliferation.

In this way, eutrophication causes ecosystem dis-equilibrium, initiating signs of pollution with ultimate loss of the value of the aquatic resources. Water is an essential natural resource for sustainability of life. It is known that a living organism requires water at some stage of its life cycle: the human, the developing embryo or foetus, the mosquito larva, the germinating plant seed or microbial spore requires water for survival. Some arid lands require only irrigation for vegetation to develop!

Indeed the abundant water on the surface of planet Earth millions of years ago, was the conducive environment for the formation of molecules into the cytoplasmic broth that constituted the precursors of the first living things.

These led to the evolution of the primitive cell (protocaryote) or prokaryote in which the nucleus is naked. Is it not interesting that bacteria and bluegreen algae belong to this group? The eucaryotic cell of fungi, protozoa and indeed multicellular organisms, including man, evolved later with a nuclear membrane and more complex organisation. It is noteworthy that the biochemical activities of microorganisms contributed a lot to the maintenance of an oxygen atmosphere responsible for the existence of life on earth.

Thus, microorganisms evolved in water and to date various species remain in their habitat of origin occupying different niches in the aquatic environment. Microbes abound in surface waters: fresh, brackish and marine. They occur in rivers, lakes, estuaries and the seas. By their microscopic size and versatile metabolic capabilities, microbes find niches and can colonize the biosphere. The number of microorganisms in surface waters is high because of the input from natural, and anthropogenic (manmade) sources including various industries. During rainfall, natural run-off washes soil microorganisms into rivers and streams that become turbid.

Also, urbanization creates large populations that generate wastes e.g. refuse (solid) and sewage (liquid). The disposal of these waste materials remains a colossal problem. There is the urbanization of farm animals and general agricultural practices including irrigation of crops that also contribute to increased number of microorganisms in water.

Bacteria are found in the shores and depths of the seas including the hypersaline dead sea with salinity above 35%o (parts per thousand). These are the extremely halophilic bacteria e.g. *Halobacterium halobium* (Haruhico *et al.*(1995). Research on halophiles has contributed to understanding the activities of osmophilic bacteria in the preservation of food with salt or sugar. Microbes get adapted to environmental changes or challenges by the modification of their metabolic pathways and evolution of genetic apparatus required to meet the demands of the new situation. This knowledge is exploited by man in the production of microbial metabolites like antibiotics, and in genetic engineering of microbes for specific characters.

The isolation of microbes from aquatic thermal vents opened a new vista for producing industrial enzymes that are stable at high temperatures (100?C). Therefore, denaturation of enzymes is avoided and the cost of cooling to protect mesophilic enzymes is saved. Eveleigh and others at Cook College of Rutgers University, USA elucidated the enzyme of an extremophilic bacterium. I had the privilege of working with Douglas Eveleigh during my last sabbatical leave.

Water also occurs underground in water tables or aquifers having percolated through the soil layers. Hence, spring water arising from the aquifer is ordinarily clear and palatable. However, seepage of effluents in refuse dumps and other human activities involving the subsurface of the earth aggravate the contamination of ground water thus interfering with the quality of spring water. Microorganisms in ground water constitute a health hazard to consumers. In a setting with little or irregular supply of pipe-borne water and usually, well water quality is not tested, consumers are vulnerable to waterborne infections.

It is the scarcity of potable water that has enhanced the sale of the so-called 'pure 'water sachets. For safety of packaged drinking water in Nigeria, the activities of the National Agency for Food Drug Administration Control (NAFDAC) deserve commendation. As an environmentalist, I would add that the used sachets constitute a nuisance especially as the waste is recalcitrant, and being non-biodegradable, it is persistent in the environment. Furthermore, care needs to be taken about the location of disposal sites for solid wastes.

The geology of the underlying soil should be considered and the disposal site may be lined before dumping, and the refuse covered with an impervious material. These efforts would control wetting of refuse by rainfall and mitigate the seepage of effluents into the aquifer with its eventual contamination and negative impact on consumers.

Water Pollution

The aquatic environment has a natural ability for self-purification. Physicochemical and biologic activities including microbiologic action bring about the breakdown of small quantities of organic materials. In effect, contaminated body of water such as a stream can be cleansed through selfpurification processes (Akpata, 1998).

It is when the type of waste is nonbiodegradable e.g. plastics or the amount of organic input exceeds the carrying capacity of the environment that there emanate undesirable conditions that elicit negative impacts on the ecosystem. This is the stage that there is environmental pollution.

Faecal Pollution of Lagos Lagoon

A trip using the famous UNILAG 1, motorboat of the erstwhile Department of Biological Sciences, was undertaken to survey the Lagos lagoon for areas of point source pollution. Nobody could ignore Iddo site of faeces disposal for the stench and unaesthetic value. There was a need to enquire from the Lagos City Council just what was going on? Iddo is located on the Lagos Mainland and is accessible to Mushin, Yaba, Ebute- Metta, Surulere, and so on. The faeces point was near the Carter Bridge on the mainland end. The rationale for disposal of raw faeces into the Lagos lagoon at Iddo was that the waste would be washed into the sea and diluted to extinction. In that case, the waste would not be seen and there would be no visible or apparent impact on the aquatic ecosystem. To clarify the situation, a research study was designed to investigate the distribution of faeces in the lagoon. Samples of fresh faeces obtained from the disposal vehicle or lagoon water were cultured for pathogens and commensals. The Lagos City Council recorded the volumes of faeces deposited into the lagoon. Seven sampling stations were identified on the lagoon, some were located to the north while others were to the south of the faeces disposal point. An apparently unpolluted station located in the middle of the lagoon served as control. Some physicochemical parameters e.g. temperature, pH, salinity, suspended organic matter and turbidity were measured at each sampling station. Faecal pollution was assayed using faecal indicator bacteria namely, coliforms, *Escherichia coli, Streptococcus faecalis or Clostridium perfringes* for their value in determining the period of pollution. Vertical and horizontal distribution of faecal bacteria in the lagoon was studied at each sampling station so as to determine the abundance and distribution of faeces in the aquatic ecosystem.

The results showed presence of pathogens e.g. *Salmonella* and *Shigella* species in the faeces samples from some areas. However, pathogens were not isolated from stations on the lagoon. It was observed that *Escherichia coli* count decreased rapidly southwards near the sea and less rapidly northwards nearer Ogun River (Akpata & Ekundayo, 1978). It is probable that stations where faecal bacteria were isolated had been contaminated with faeces and possibly pathogens. There was both horizontal and vertical distribution of faeces in the lagoon. Higher counts of bacteria were obtained in bottom samples an indication that sedimentation of faeces occurred at each station. This could be due to the effect of gravity and increased organic nutrients. Twenty-six million litres of faeces were deposited into the lagoon at Iddo in 1973. Our results showed that the faeces were not always washed southwards into the sea from the Lagos lagoon as suggested by the Lagos City Council.

WATER MICROBIOLOGY

Water microbiology is concerned with the microorganisms that live in water, or can be transported from one habitat to another by water.

Water can support the growth of many types of microorganisms. This can be advantageous. For example, the chemical activities of certain strains of yeasts provide us with beer and bread. As well, the growth of some bacteria in contaminated water can help digest the poisons from the water.

However, the presence of other disease causing microbes in water is unhealthy and even life threatening. For example, bacteria that live in the intestinal tracts of humans and other warm blooded animals, such as *Escherichia coli, Salmonella, Shigella,* and *Vibrio,* can contaminate water if feces enters the water. Contamination of drinking water with a type of *Escherichia coli* known as O157:H7 can be fatal. The contamination of the municipal water supply of Walkerton, Ontario, Canada in the summer of 2000 by strain O157:H7 sickened 2,000 people and killed seven people.

The intestinal tract of warm-blooded animals also contains viruses that can contaminate water and cause disease. Examples include rotavirus, enteroviruses, and coxsackievirus.

Another group of microbes of concern in water microbiology are protozoa. The two protozoa of the most concern are *Giardia* and *Cryptosporidium*. They live normally in the intestinal tract of animals such as beaver and deer. *Giardia* and *Cryptosporidium* form dormant and hardy forms called cysts during their life cycles. The cyst forms are resistant to chlorine, which is the most popular form of drinking water disinfection, and can pass through the filters used in many water treatment plants. If ingested in drinking water they can cause debilitating and prolonged diarrhea in humans, and can be life threatening to those people with impaired immune systems. *Cryptosporidium* contamination of the drinking water of Milwaukee, Wisconsin with in 1993 sickened more than 400,000 people and killed 47 people. Many microorganisms are found naturally in fresh and saltwater. These include bacteria, cyanobacteria, protozoa, algae, and tiny animals such as rotifers. These can be important in the food chain that forms the basis of life in the water. For example, the microbes called cyanobacteria can convert the energy of the sun into the energy it needs to live. The plentiful numbers of these organisms in turn are used as food for other life. The algae that thrive in water is also an important food source for other forms of life. A variety of microorganisms live in fresh water. The region of a water body near the shoreline (the littoral zone) is well lighted, shallow, and warmer than other regions of the water. Photosynthetic algae and bacteria that use light as energy thrive in this zone. Further away from the shore is the limnitic zone. Photosynthetic microbes also live here. As the water deepens, temperatures become colder and the oxygen concentration and light in the water decrease. Now, microbes that require oxygen do not thrive. Instead, purple and green sulphur bacteria, which can grow without oxygen, dominate. Finally, at the bottom of fresh waters (the benthic zone), few microbes survive. Bacteria that can survive in the absence of oxygen and sunlight, such as methane producing bacteria, thrive.

Saltwater presents a different environment to microorganisms. The higher salt concentration, higher pH, and lower nutrients, relative to freshwater, are lethal to many microorganisms. But, salt loving (halophilic) bacteria abound near the surface, and some bacteria that also live in freshwater are plentiful (i.e., *Pseudomonas* and *Vibrio*). Also, in 2001, researchers demonstrated that the ancient form of microbial life known as archaebacteria is one of the dominant forms of life in the ocean. The role of archaebacteria in the ocean food chain is not yet known, but must be of vital importance. Another microorganism found in saltwater are a type of algae known as dinoflagellelates. The rapid growth and multiplication of dinoflagellates can turn the water red. This "red tide" depletes the water of nutrients and oxygen, which can cause many fish to die. As well, humans can become ill by eating contaminated fish. Water can also be an ideal means of transporting microorganisms from one place to another. For example, the water that is carried in the hulls of ships to stabilize the vessels during their ocean voyages is now known to be a means of transporting microorganisms around the globe. One of these organisms, a bacterium called *Vibrio cholerae*, causes life threatening diarrhea in humans. Drinking water is usually treated to minimize the risk of microbial contamination. The importance of drinking water treatment has been known for centuries. For example, in pre-Christian times the storage of drinking water in jugs made of metal was practiced. Now, the anti-bacterial effect of some metals is known. Similarly, the boiling of drinking water, as a means of protection of water has long been known. Chemicals such as chlorine or chlorine derivatives has been a popular means of killing bacteria such as *Escherichia coli* in water since the early decades of the twentieth century. Other bacteria-killing treatments that are increasingly becoming popular include the use of a gas called ozone and the disabling of the microbe's genetic material by the use of ultraviolet light. Microbes can also be physically excluded form the water by passing the water through a filter. Modern filters have holes in them that are so tiny that even particles as miniscule as viruses can be trapped.

An important aspect of water microbiology, particularly for drinking water, is the testing of the water to ensure that it is safe to drink. Water quality testing can de done in several ways. One popular test measures the turbidity of the water. Turbidity gives an indication of the amount of suspended material in the water. Typically, if material such as soil is present in the water then microorganisms will also be present. The presence of particles even as small as bacteria and viruses can decrease the clarity of the water. Turbidity is a quick way of indicating if water quality is deteriorating, and so if action should be taken to correct the water problem. In many countries, water microbiology is also the subject of legislation. Regulations specify how often water sources are sampled, how the sampling is done, how the analysis will be performed, what microbes are detected, and the acceptable limits for the target microorganisms in the water sample. Testing for microbes that cause disease (i.e., *Salmonella typhymurium* and *Vibrio cholerae*) can be expensive and, if the bacteria are present in low numbers, they may escape detection. Instead, other more numerous bacteria provide an indication of fecal pollution of the water. *Escherichia coli* has been used as an indicator of fecal pollution for decades. The bacterium is present in the intestinal tract in huge numbers, and is more numerous than the disease-causing bacteria and viruses. The chances of detecting *Escherichia coli* is better than detecting the actual disease causing microorganisms. *Escherichia coli* also had the advantage of not being capable of growing and reproducing in the water (except in the warm and food-laden waters of tropical countries). Thus, the presence of the bacterium in water is indicative of recent fecal pollution. Finally, *Escherichia coli* can be detected easily and inexpensively.

There are two major types of water.

1. Ground water - It originates from deep wells and subterranean springs. This is virtually free of bacteria due to filtering action of soil, deep sand and rock. However, it may become contaminated when it flows along the channels.
2. Surface water - It is found in streams, lakes, and shallow wells. The air through which the rain passes may contaminate the water. Other sources are the various types of establishments and agricultural farms etc. by the sides the water flows.

Possible sources of microbial contamination of a body of water are soil and agricultural run off, farm animals, rain water, industrial waste, discharges from sewage treatment plants and storm water run off from urban areas. In water microbiology the water is contaminated when it contains a chemical or biological poison or an infectious agent. These conditions also apply to water which is polluted except that the agent or poison is often obvious and the water carries an unpleasant taste or appearance. Potability refers to the drinkability of water. When potable, it is fit for drinking. When unpotable it is unfit due to some contaminant or pollutant.

ATMOSPHERIC WATERS

Rain, snow and hail which fan on land tend to carry down particles of dust, soot, and other materials suspended in the air. These often bear bacteria and other microorganisms -on their surface. The number of organisms depends upon local conditions. After heavy rain or snow the atmosphere is washed free of organisms. Surface Waters - As soon as ram or snow reaches the earth and flows over the soil, some of the soil organisms are gathered up by the water. Bodies, of water such as streams, rivers, and oceans represent surface water.

Microbial populations depend upon their numbers in the soil and, also, upon the kinds and quantities of food material dissolved out of the soil by water. Climatic, geographical and biological conditions bring about great variations in microbial populations of surface waters.

Rivers and streams show their highest count during the rainy period Dust blowing into rivers and streams also contributes many microorganisms. Animals also make considerable contribution to the microbial flora of the surface waters. They bathe and often drop their excreta in the water.

Stored Waters - Inland waters held in ponds, lakes or reservoirs represent stored waters. Storage generally reduces the numbers of organisms in water. A certain degree (If purity and stability is established.

Several factors affect the microbial flora of stored waters. These are as follows;

Sedimentation

Microorganisms have a specific gravity slightly greater than that of' water, and therefore slowly settle down. However, the most important factor is their attachment to suspended particles. Microorganisms are removed from the upper layers of the water as the suspended particles settle down.

Activities of other Organisms

Predatory Protozoa engulf living or dead bacteria for food, provided the water contains sufficient dissolved oxygen.

Light ray: Direct sunlight is toxic to both vegetative cells and spores of microorganisms.

The toxicity of ultraviolet rays is inversely proportional to the turbidity of water. In tropical countries direct sun light is a very effective sterilizing agent.

Temperature

Temperature has variable effects. It may kill some organisms and may stimulate the growth of others. During colder months the multiplication rate of microorganisms is considerably reduced.

Food Supply

If there is considerable vegatation or suspended food particles in the body of water, it is likely to increase the number of organisms. On the other hand, certain toxic substances may bring about marked reduction in the number of, organisms.

Ground Waters

As water seeps through the earth, microorganisms as well as suspended particles are removed by filtration in varying degrees. This depends on the permeability characteristics of the soil and the depth to which the water penetrates.

Ground water brought to the surface by springs or deep wells contains very few organisms. To construct a well the nature of the soil and underlying porous strata, the nature of the water table, and the nature, distance, and direction of the local sources of pollution must be taken into consideration.

Faecal Contamination of Water

Water gets contaminated with pathogenic microorganisms through intestinal discharges of man and animals. Further more, in the intestinal tract of man and animals, there exists a characteristic group of organism's designated as coliforms.

The coliform group of bacteria includes aerobic and facultatively anaerobic, gram negative, nonsporeforming bacilli which ferment lactose with acid and produce gas within 48 hours at, 35°C. The most common species of this group are various strains of Escherichia coli and Aerobacter aerogenes.

E. coli is commonly found in the intestinal tract of man and animals, while A. aerogenes is normally found on plants and grains, and may sometimes occur in the intestinal tract of man and animals. Contamination of water with either type makes the water unsatisfactory for drinking purposes.

Marine Microbiology

The study of microorganisms living in the sea is known as marine microbiology. Nearly three fourths of the earth's surface is covered by the sea.The sea therefore is the largest natural environment inhabited by the microorganisms. The marine environment, as does land, contains bacteria, Protozoa, algae, yeasts and moulds, as well as viruses forms organisms that live in the sea.

Microorganisms which are found in the surface layers of the sea and other aquatic environment are collectively known as plankton.

Plankton may be further sub divided into phytoplank-ton (photosynthetic microorganisms, algae) and zooplankton (Protozoa, and other microscopic animals). The former are more important plankton organisms, as they arc the primary producers of organic matter.

Bacteria and fungi are also present in large numbers in the plankton. Microorganisms present in the bottom regions (sediments) arc designated the benthos or benthic community. A variety of microorganisms arc found in this region. Bacteria, however, predominate.

The microbial flora characteristic of tile sea has one factor in its environmental pattern which differs from freshwater lakes, streams, and rivers, for the sea is salty. There is a high concentration of salt and mineral ions in the sea. The principal salts are the chlorides, sulphates, and carbonates of sodium, potassium, calcium and magnesium.

The concentration of dissolved salts (salinity) averages about 35 gin/kg or 3.5% by weight. Out of this 75% is NaCl. Marine microorganisms, there-fore; may be divided into two general types:

(1) those indigenous to the sea and not growing on media without sea water or high salt concentra-tion and
(2) transient organisms whose natural habitat is terrestrial and which are able to grow in media without sea water, but can tolerate high salt concentration.

There is also another group of microorganisms in the sea, in the depths of the sea microorganisms live at tremendous hydrostatic pressures, upto 1000 atmospheres. These organisms when brought to the surface require not only sea water but high pressure in order to grow. Such pressures are toxic to organisms normally living at one atmosphere at the surface.

Bacteriological Examination of Domestic Water

Natural water supplies such as rivers, lakes, and streams contain sufficient nutrients to support growth of various organisms. Microorganisms enter the water supply in several different ways. In congested centers water supplies get polluted by domestic and industrial wastes, As a potential carrier of pathogenic microorganisms, water can endanger health and life.

From the standpoint of transmitting human diseases, polluting waters with soil, rubbish, industrial wastes, and even animal manure is comparatively harmless. These sources rarely contain pathogens capable of producing human diseases when swallowed with drinking water. Sewage containing human excreta, however, is the most dangerous material that pollutes water.

People with communicable diseases of many kinds eliminate the causative organisms in their excreta. The most important microbial diseases transmitted through water are typhoid fever, paratyp-hoid fever, amoebic dysentery, bacillary dysentery, cholera, tularemia, poliomyelitis, and infectious hepatitis.

To determine the potability of water quantitative bacteriological examination may be undertaken. However, there is no single test or, even combination of tests, that is wholly satisfactory, because it will give only a fraction of the total count. Theoretically it would be better to examine water for the presence of the specific pathogenic microorganisms. This is also impracticable because of the following reasons:

1. The methods are expensive, tedious, and slow, and by that time the water has already been consumed.
2. The number of pathogenic organisms may be quite small compared to non pathogenic organisms and would be overlooked.
3. Non-pathogenic organisms may interfere with the examination of pathogens.

The direct examination for pathogens, therefore, is not used in routine water analysis. Methods commonly used for the bacteriological examination of water, are based on:

1. The examination of presence or absence of the more common organisms of intestinal or sewage origin.
2. The approximate determination of total numbers of bacteria present in the water sample.

Microbial Components of Water

The type of microorganisms differ in unpolluted, polluted and marine waters.. In unpolluted water of mountain lake or stream there are usually low organic nutrients.

The number of bacteria is very much limited, a few thousand per ml. Actinomycetes are typical. Autotrophic bacteria are common alongwith free living protozoans as Euglena, Paramecium and various amoebae.

Marine Water

The marine water is high in salt and only halophilic organisms survive. Since temperature is very low, most of them are psychrophilic. Diatoms and protozoa as dinoflagellates are important components in food chain. In the off shore oceanic zone, photosynthetic organisms are present. These are diatoms and dinoflagellates.

Most marine microbes are found along the shoreline or littoral zone where nutrients are in abundance. Some unusual types are also present on the ocean floor in the benthic zone and even at the bottom of several miles-deep trenches, the abyssal zone.

Polluted Water

In polluted waters, there are large amounts of organic matter from sewage, feces and industrial complex. The microbes are usually heterotrophic. The digestion of organic matter by these organisms is incomplete, due to which there accumulate acids, bases, alcohols and various gases.

The major type of bacteria are coliform bacteria the Gram-negative nonsporeforming bacilli usually found in the intestine. This group includes E. coli and species of Enterabacter. They ferment lactose to acid and gas. Noncoliform bacteria Streptococcus, Proteus and Pseudomonas are also present.

Under some conditions, the polluting organisms multiply rapidly and consume most of the available oxygen. For instance, nutrients enter the river from sources like sewage treatment plants or urban suburban runoff.

Thus river suddenly develops a high nutrient content. Under these conditions algae may bloom rapidly. This leads to depletion of oxygen in water. There is very little oxygen available to the protozoa, small animals, fish and plants. Due to this non-availability of oxygen, a layer of dead organisms, mud and silt accumulate at the bottom, and anaerobeic species of Clostridium, Desulfovibrio etc. will flourish.

They produce gases. One gas, H2S combines with lead or iron to give a precipitate which makes the mud black and the water poisonous. Due to complete depletion of oxygen, the suspended bacteria die in their own waste products. There is hardly any life in water at this stage. The gas bubbles from the anaerobes in the mud break the surface. Such processes lead to death of a river.

Water Pollution

Polluted waters can be classified in different ways. In microbiological context, there are generally recognised physical, chemical and biological pollution of water.

Physical Water Pollution

Here the water becomes cloudy due to presence of some particulate matter such as sand or soil. Some plant material as algal-bloom may also be involved, that turns the consistency of water like a pea soup. Presence of some phosphates or nitrates may encourage the growth of blooms leading finally to a condition known as eutrophication. This upsets the ecological balance.

Chemical Water Pollution

It occurs due to introduction of inorganic and organic waste to water. For example, water passing out of a mine contains large amount of copper, iron or other ore. Another source of pollution is metal pipelines. Corrosion of metals of the pipes may yield rusty or black precipitates.

Other chemical pollutants include laundry detergents, radioactive wastes and nitrates from fertilisers etc. An important group of chemicals are polychlorinated biphenyls or PCBs, a group of industrial chemicals used in the production of electric insulators and paper products.

They are carcinogenic and necessary to be eliminated from water. Main sources of chemical pollutants are mining wastes, air pollutants (washed down by rain), power plants, industries, sewage and storm water runoff, wash from farm lands (pesticides, fertilisers), hospitals, road salts and exhausts from automobiles.

Biological Water Pollution

It develops from the microorganisms that enter waters from such sources as human waste, food operations, meat packing plants and medical facilities etc. Under normal conditions, the water body is able to handle biological material, because heterotrophs digest the organic matter to simple phosphates, nitrates and sulphates which fertilise the plants.

Carbohydrates are converted to CO2 and water, which are also useful to plants. The plants then release °2 through photosynthesis. Due to rapid movement of water, aeration is more or less constant and the wastes eliminated.

However, there arises the problem when water becomes stagnant or is overloaded with waste. Water is now unable to handle the biological material, and it soon becomes polluted. We have already seen how a lake may turn into a swamp due to overabundance of organic matter.

A critical factor operating under such a situation is the biological oxygen demand (BOD). It refers to the oxygen requirement by the metabolising organisms. As the number of organisms increases, the demand for O2 increases proportionally. Depletion of O2 in water may lead to death of fish and other aquatic animals, a flat taste of water, and eventual death of aerobes.

The BOD is therefore an important indication of the levels of biological pollution in the water. In laboratory, BOD is determined by measuring the amount of dissolved oxygen in test samples of water distributed to BOD bottles. Initial determination of dissolved oxygen in sample is made.

The bottles are then stoppered and incubated at 20°C for five days, after which the amount of dissolved oxygen is again determined. The difference in the oxygen content represents the amount of oxygen used up in the test sample, which is the BOD. Results are expressed as parts per million (ppm). A BOD of several hundred is taken usually high.

Diseases Transmitted by Water

Water is infact a vehicle for the transfer of a wide range of diseases of microbial origin. The important bacterial diseases include typhoid fever, cholera and bacterial dysentry that are generally transmitted when human feces from carriers or patients contaminates the water.

Viral diseases transmitted by water include virus A hepatitis and polio. Many protozoa form cysts which survive for long periods in water. The causal agents of amoebiasis, giardiasis etc. are important concerns in water pollution.

Water Treatment

We have seen that water and sewage harbour a variety of organisms responsible for diseases in man. It is thus necessary to remove or rather interrupt the disease cycle in nature through treatment of water and sewage. Water purification and sanitary sewage disposal are two major methods for interrupting disease cycle. In water purification organisms are prevented from reaching the body whereas in sewage disposal, they are removed from body waste products. The various steps in the purification of drinking water in municipal water supplies are, It may be seen that there are three basic steps in the purification of drinking water: sedimentation, filtration and chlorination.

Sedimentation Process

Water usually undergoes some degree of purification during storage in ponds or reservoirs. Suspended particles settle and carry down most of the microorganisms. The rate of purification by sedimentation depends upon the kind and amount of suspended matter as the well as physical, chemical and biological conditions of the stored water. The rate of sedimentation is enhanced by adding alum, iron salts, colloidal silicate, etc., which produce flocculent precipitates. Microorganisms and suspended particles are entrapped and settle rapidly. Sometimes activated carbon is also added. This adsorbs the compound responsible for objectionable colour and taste of water. Microorganisms remain viable for a considerable time, even though visible evidence of pollution has disappeared. Sedimentation, therefore, reduces the microbial population but does not sterilize polluted water. To produce potable water further treatment is necessary. Thus sedimentation is often used as a first stage purification.

Filtration Process

This process removes the remaining microorganisms from the water. Although, different types of filter material are available, mostly a layer of sand and gravel is utilised to trap the microorganisms.

A slow sand filter uses several feet deep layer of fine sand particles. Within the sand there forms a layer of microorganisms which acts as an additional filter. This layer is called a Schmutzdecke or dirty layer.

To cleanse the filter, the top layer is removed and replaced with fresh sand. A rapid sand filter contains coarser particles of gravel. The schmutzdecke does not develop; the filtration rate is higher (over 200 million gallons per acre per day). This type of filter is generally used in municipal supplies.

Slow Sand Filter

Slow sand filtration plants require considerable area because the rate of filtration is slow. A concrete floor with drainage tiles to collect the filtered water is constructed. The tile is covered with coarse gravel, fine gravel, coarse sand and finally 2 to 1 feet of sand at the top. Water seeps through the filter slowly., is collected by tile drain pipes at the bottom, and is pumped into a reservoir. At best five million gallons of water per acre, per day, can be filtered. Slow sand filters are clogged by turbid water. Water to be filtered is, therefore, clarified by sedimentation with or without coagulation. The purification of water is accomplished not by the screening action of the sand for the spaces arc much to large, but by a different principle. A colloidal, flocculent material composed of bacteria, algae, and Protozoa accumulates in the surface layers of fine sand. This Slimy, gelatinous film closes up the pores between the sand grains and makes the filter bed more and more effective. Since bacteria have a negative electrical charge and colloidal material on the sand grains has a. positive charge, bacteria are thus adsorbed on the particles.

Bacteria arc also injested by Protozoa that Inhabit the upper layer of the film. Metabolic activity of microorganisms also greatly reduces the chemical content of the water. When, the gelatinous film finally become too thick, the efficiency of the filter gradually decreases. The filter is taken out of service and the surface layer is cleaned.

Rapid Sand Filter

Rapid sand filters are constructed in a manner similar to that of slow sand filters. They also consist of layers of sand, gravel, and rock. Water is pretreated before filtration by a coagulant such as alum or ferrous sulphate.

The water passes through a settling tank in which most of the precipitate settles out, and the remainder is pumped on to the filter. Rapid sand filters soon become clogged and are cleaned by forcing cleaned water backward (back washed) through the bed of gravel and sand, and bubbling air through them.

The back water rises through the filter and carries the accumulated material to the sewer. The wash water is thus wasted, Care is taken in this backwashing procedure to see that the fine sand on the surface is not lost.

Rapid sand filters are usually operated in batteries, so that some may be in operation, while others are being cleaned.

They are nearly as effective as slow sand filters but operate 50 times faster than slow sand filters. Rapid sand filters are capable of delivering 150 to 200 million gallons of water per acre, per day. They require a much smaller area of land for more water filtrations and cost much less to install and maintain.

Many other filtration devices such as pressure filters, diatomite fitters, membrane filter, reverse osmosis etc, are employed to remove various impurities in water. Recovery of potable water from the sea and from domestic and. industrial sewage is also undertaken by the use of filtration techniques.

Chlorination Process

In this final step, chlorine gas is added to the water. This gas is an active oxidant which reacts with organic matter in the water. Gas is added until any residue is present. A residue of 0.2 ppm-1.0 ppm of water is often the standard used. Under these conditions, most remaining microbes die within 30 minutes.

Bacteriological Analysis of Water

There are methods to detect bacterial contamination of water. The chief objective is to identify coliform organisms as Escherichia coli. Their presence indicates that water contains fecal pollution and is unsafe for consumption.

Membrane Filter Technique

It may be used in the field also. A special, collecting bottle is held against the water current and a 100-ml sample is taken. The water is filtered through a membrane filter and the filter pad is then transferred to a plate of bacteriological medium. Bacteria trapped in the filter will form colonies which may be counted.

Standard Plate Count SPC

Water sample is diluted in sterile buffer. Measured amounts are pipetted into Petri dishes. Agar medium is added and plates incubated. Colony counts are made and multiplied by the reciprocal dilution factor to have total bacteria per ml water. This method is similar to that used for milk.

Most Propable Number

In this procedure, water in 10 ml, I mi and 0.1 mt amounts is inoculated into lactose broth tubes. The tubes are incubated and coliform organisms may be identified by their production of gas from the lactose.

By referring to a MPN table, a statistical range of the number of coliform may be determined by observing how many broth tubes showed gas.

This method does not detect total number of bacteria in the water nor it locate noncoliforms like Salmonella. However it indicates the presence and quantity of coliforms.

Specific Tests For Asseing Water Coliform Counts

The most frequently used indicator organism is the normally nonpathogenic Coliform bacterium Escherichia coli. Positive tests for E. coli do not prove the presence of enteropathogens, such as Salmonella and Shigella, but do establish the possibility. E. coli is more numerous and easier to be grown.

Presumptive Test

A series of lactose broth or lauryl sulphate tryptose broth fermentation tubes are inoculated with measured amounts of water and incubated at 35°C for 24 to 48 hours. The formation of gas in the inverted vial in the fermentation tube within 48 hours indicates a positive presumptive test. Absence of gas formation at the end of 48 hours constitutes a negative presumptive test. This means that the water sample does not contain coliforms and is considered safe.

Confirmed Test

Sometimes a false positive presumptive test is obtained. This may be due to the presence of yeasts, certain Clostridium species and some other organisms, or by bacterial associations or synergism. In order to be certain that gas-production is due to coliforms a confirmed test must be performed. Two procedures are normally employed. In one method a drop of culture from a positive lactose broth tube is transferred to brilliant green lactose bile fermentation broth, and is incubated for 24 to 48 hours at 35°C. The appearance of gas within 48 hrs constitues a positive confirmed test. The dye inhibits gram positive organisms and eliminates a false presumptive test and the syngergistic reaction of gram positive and gram negative organisms growing together. In the second method a drop of culture from the positive lactose broth is streaked on a petriplate containing, Endo's, or eosin-methylene blue agar. The appearance of nucleated colonies, with or without a metallic sheen, within 24 hours indicates a positive confirmed test.

Completed Test

Isolated colonies from petriplates are transferred into lactose fermentation broth and streaked on to an agar slant. The presence of gas in the fermentation broth and the presence of gram negative non spore-forming bacilli on the slant give evidence that coliform bacteria were present in the original water sample.

IMViC Reaction

E. coli and A. aerogenes are normally referred to as faecal and non-faecal contaminants of water, respectively, and are the most important organisms of the coliform group. Since they closely resemble each other in their morphological and cultural characteristics, biochemical tests, are, performed to differentiate them. These tests are collectively designated as the IMViC reactions. The name was coined by Parr from the first letters of the four tests, namely Indole, Methyl red, Voges Proskauer, and Citrate. There are 16 possible combinations of positive and negative tests of these four characteristics. Most, of these combinations have been found, but the reactions of E. coli and A. aerogenes are commonly found. The remaining 14 types are usually designated as "intermediates".

ENVIRONMENTAL BIOSPHERE

The biosphere is the region of the earth that encompasses all living organisms: plants, animals and bacteria. It is a feature that distinguishes the earth from the other planets in the solar system. "Bio" means life, and the term biosphere was first coined by a Russian scientist (Vladimir Vernadsky) in the 1920s. Another term sometimes used is ecosphere ("eco" meaning home). The biosphere includes the outer region of the earth (the lithosphere) and the lower region of the atmosphere (the troposphere).

It also includes the hydrosphere, the region of lakes, oceans, streams, ice and clouds comprising the earth's water resources. Traditionally, the biosphere is considered to extend from the bottom of the oceans to the highest mountaintops, a layer with an average thickness of about 20 kilometers. Scientists now know that some forms of microbes live at great depths, sometimes several thousand meters into the earth's crust.

Nonetheless, the biosphere is a very tiny region on the scale of the whole earth, analogous to the thickness of the skin on an apple. The bulk of living organisms actually live within a smaller fraction of the biosphere, from about 500 meters below the ocean's surface to about 6 kilometers above sea level.

Dynamic interactions occur between the biotic region (biosphere) and the abiotic regions (atmosphere, lithosphere and hydrosphere) of the earth. Energy, water, gases and nutrients are exchanged between the regions on various spatial and time scales. Such exchanges depend upon, and can be altered by, the environments of the regions. For example, the chemical processes of early life on earth (e.g. photosynthesis, respiration, carbonate formation) transformed the reducing ancient atmosphere into the oxidizing (free oxygen) environment of today. The interactive processes between the biosphere and the abiotic regions work to maintain a kind of planetary equilibrium. These processes, as well as those that might disrupt this equilibrium, involve a range of scientific and socioeconomic issues. The study of the relationships of living organisms with one another and with their environment is the science known as ecology. The word ecology comes from the Greek words oikos and logos, and literally means "study of the home." The ecology of the earth can be studied at various levels: an individual (organism), a population, a community, an ecosystem, a biome or the entire biosphere. The variety of living organisms that inhabit an environment is a measure of its biodiversity.

ORGANISMS

Life evolved after oceans formed, as the ocean environment provided the necessary nutrients and support medium for the initial simple organisms. It also protected them from the harsh atmospheric UV radiation. As organisms became more complex they eventually became capable of living on land. However, this could not occur until the atmosphere became oxidizing and a protective ozone layer formed which blocked the harmful UV radiation. Over roughly the last four billion years, organisms have diversified and adapted to all kinds of environments, from the icy regions near the poles to the warm tropics near the equator, and from deep in the rocky crust of the earth to the upper reaches of the troposphere. Despite their diversity, all living organisms share certain characteristics: they all replicate and all use DNA to accomplish the replication process. Based on the structure of their cells, organisms can be classified into two types: eukaryotes and prokaryotes. The main difference between them is that a eukaryote has a nucleus, which contains its DNA, while a prokaryote does not have a nucleus, but instead its DNA is free-floating in the cell. Bacteria are prokaryotes, and humans are eukaryotes. Organisms can also be classified according to how they acquire energy. Autotrophs are "self feeders" that use light or chemical energy to make food. Plants are autotrophs. Heterotrophs (i.e. other feeders) obtain energy by eating other organisms, or their remains. Bacteria and animals are heterotrophs. Groups of organisms that are physically and genetically related can be classified into species. There are millions of species on the earth, most of them unstudied and many of them unknown. Insects and microorganisms comprise the majority of species, while humans and other mammals comprise only a tiny fraction. In an ecological study, a single member of a species or organism is known as an individual.

POPULATIONS AND COMMUNITIES

A number of individuals of the same species in a given area constitute a population. The number typically ranges anywhere from a few individuals to several thousand individuals. Bacterial populations can number in the millions. Populations live in a place or environment called a habitat. All of the populations of species in a given region together make up a community. In an area of tropical grassland, a community might be made up of grasses, shrubs, insects, rodents and various species of hoofed mammals. The populations and communities found in a particular environment are determined by abiotic and biotic limiting factors. These are the factors that most affect the success of populations. Abiotic limiting factors involve the physical and chemical characteristics of the environment. Some of these factors include: amounts of sunlight, annual rainfall, available nutrients, oxygen levels and temperature.

For example, the amount of annual rainfall may determine whether a region is a grassland or forest, which in turn, affects the types of animals living there. Each population in a community has a range of tolerance for an abiotic limiting factor. There are also certain maximum and minimum requirements known as tolerance limits, above and below which no member of a population is able to survive. The range of an abiotic factor that results in the largest population of a species is known as the optimum range for that factor. Some populations may have a narrow range of tolerance for one factor. For example, a freshwater fish species may have a narrow tolerance range for dissolved oxygen in the water. If the lake in which that fish species lives undergoes eutrophication, the species will die. This fish species can therefore act as an indicator species, because its presence or absence is a strict indicator of the condition of the lake with regard to dissolved oxygen content.

Biotic limiting factors involve interactions between different populations, such as competition for food and habitat. For example, an increase in the population of a meat-eating predator might result in a decrease in the population of its plant-eating prey, which in turn might result in an increase in the plant population the prey feeds on. Sometimes, the presence of a certain species may significantly affect the community make up. Such a species is known as a keystone species. For example, a beaver builds a dam on a stream and causes the meadow behind it to flood. A starfish keeps mussels from dominating a rocky beach, thereby allowing many other species to exist there.

Ecosystems

An ecosystem is a community of living organisms interacting with each other and their environment. Ecosystems occur in all sizes. A tidal pool, a pond, a river, an alpine meadow and an oak forest are all examples of ecosystems. Organisms living in a particular ecosystem are adapted to the prevailing abiotic and biotic conditions. Abiotic conditions involve both physical and chemical factors (e.g., sunlight, water, temperature, soil, prevailing wind, latitude and elevation). In order to understand the flow of energy and matter within an ecosystem, it is necessary to study the feeding relationships of the living organisms within it.

Living organisms in an ecosystem are usually grouped according to how they obtain food. Autotrophs that make their own food are known as producers, while heterotrophs that eat other organisms, living or dead, are known as consumers. The producers include land and aquatic plants, algae and microscopic phytoplankton in the ocean. They all make their own food by using chemicals and energy sources from their environment.

For example, plants use photosynthesis to manufacture sugar (glucose) from carbon dioxide and water. Using this sugar and other nutrients (e.g., nitrogen, phosphorus) assimilated by their roots, plants produce a variety of organic materials. These materials include: starches, lipids, proteins and nucleic acids. Energy from sunlight is thus fixed as food used by themselves and by consumers.

The consumers are classed into different groups depending on the source of their food. Herbivores (e.g. deer, squirrels) feed on plants and are known as primary consumers. Carnivores (e.g. lions, hawks, killer whales) feed on other consumers and can be classified as secondary consumers. They feed on primary consumers. Tertiary consumers feed on other carnivores. Some organisms known as omnivores (e.g., bears, rats and humans) feed on both plants and animals. Organisms that feed on dead organisms are called scavengers (e.g., vultures, ants and flies). Detritivores (detritus feeders, e.g. earthworms, termites, crabs) feed on organic wastes or fragments of dead organisms.

Decomposers (e.g. bacteria, fungi) also feed on organic waste and dead organisms, but they digest the materials outside their bodies. The decomposers play a crucial role in recycling nutrients, as they reduce complex organic matter into inorganic nutrients that can be used by producers. If an organic substance can be broken down by decomposers, it is called biodegradable.

In every ecosystem, each consumer level depends upon lower-level organisms (e.g. a primary consumer depends upon a producer, a secondary consumer depends upon a primary consumer and a tertiary consumer depends upon a secondary consumer).

All of these levels, from producer to tertiary consumer, form what is known as a food chain. A community has many food chains that are interwoven into a complex food web. The amount of organic material in a food web is referred to as its biomass.

When one organism eats another, chemical energy stored in biomass is transferred from one level of the food chain to the next. Most of the consumed biomass is not converted into biomass of the consumer. Only a small portion of the useable energy is actually transferred to the next level, typically 10 percent. Each higher level of the food chain represents a cumulative loss of useable energy. The result is a pyramid of energy flow, with producers forming the base level. Assuming 10 percent efficiency at each level, the tertiary consumer level would use only 0.1 percent of the energy available at the initial producer level. Because there is less energy available high on the energy pyramid, there are fewer top-level consumers. A disruption of the producer base of a food chain, therefore, has its greatest effect on the top-level consumer.

Ecosystem populations constantly fluctuate in response to changes in the environment, such as rainfall, mean temperature, and available sunlight.

Normally, such changes are not drastic enough to significantly alter ecosystems, but catastrophic events such as floods, fires and volcanoes can devastate communities and ecosystems. It may be long after such a catastrophic event before a new, mature ecosystem can become established. After severe disturbance the make up of a community is changed. The resulting community of species changes, as early, post disturbance, fast-growing species are out-competed by other species. This natural process is called ecological succession. It involves two types of succession: primary succession and secondary succession. Primary succession is the development of the first biota in a given region where no life is found. An example is of this is the surrounding areas where volcanic lava has completely covered a region or has built up a new island in the ocean. Initially, only pioneer species can survive there, typically lichens and mosses, which are able to withstand poor conditions.

They are able to survive in highly exposed areas with limited water and nutrients. Lichen, which is made up of both a fungus and an alga, survives by mutualism. The fungus produces an acid, which acts to further dissolve the barren rock. The alga uses those exposed nutrients, along with photosynthesis, to produce food for both. Grass seeds may land in the cracks, carried by wind or birds. The grass grows, further cracking the rocks, and upon completing its own life cycle, contributes organic matter to the crumbling rock to make soil. In time, larger plants, such as shrubs and trees may inhabit the area, offering habitats and niches to immigrating animal life. When the maximum biota that the ecosystem can support is reached, the climax community prevails. This occurs after hundreds if not thousands of years depending on the climate and location.

Secondary succession begins at a different point, when an existing ecosystem's community of species is removed by fire, deforestation, or a bulldozer's work in a vacant lot, leaving only soil. The first few centimeters of this soil may have taken 1000 years to develop from solid rock.

It may be rich in humus, organic waste, and may be stocked with ready seeds of future plants. Secondary succession is also a new beginning, but one with a much quicker regrowth of organisms. Depending on the environment, succession to a climax community may only require 100 to 200 years with normal climate conditions, with communities progressing through stages of early plant and animal species, mid-species and late successional species. Some ecosystems, however, can never by regained.

The grass grows, further cracking the rocks and upon completing its own life cycle, contributes organic content to the crumbling rock becoming soil. In time, larger plants then shrubs then trees may dominate this area, offering habitats and niches to immigrating animal life. Reaching the maximum biota this ecosystem can support, the climax community caps off further succession after hundreds if not thousands of years depending on the climate and location. Secondary succession begins at a different point. Due to a disturbance to an existing ecosystem, such as fire, deforestation, farmland left abandoned, or even the bulldozer's work in a vacant lot, biota are removed. What does remain, however, is the soil. Soil that may have taken 1000 years to develop the first centimeter from solid rock. Soil that may be rich in humus, organic wastes. Soil that may be stocked with ready seeds of future plants. Secondary succession is also new beginning, but with the advantage of quick regrowth of organisms.

Once underway, the ecosystem's succession to its climax community may only require 100-200 years with normal climate conditions, progressing just like the primary succession through stages of early plant and animal species, mid-species, and late successional species.

Biomes

The biosphere can be divided into relatively large regions called biomes. A biome has a distinct climate and certain living organisms (especially vegetation) characteristic to the region and may contain many ecosystems. The key factors determining climate are average annual precipitation and temperature. These factors, in turn, depend on the geography of the region, such as the latitude and elevation of the region, and mountainous barriers. The major types of biomes include: aquatic, desert, forest, grassland and tundra. Biomes have no distinct boundaries. Instead, there is a transition zone called an ecotone, which contains a variety of plants and animals. For example, an ecotone might be a transition region between a grassland and a desert, with species from both.

Water covers a major portion of the earth's surface, so aquatic biomes contain a rich diversity of plants and animals. Aquatic biomes can be subdivided into two basic types: freshwater and marine. Freshwater has a low salt concentration, usually less than 1 percent, and occurs in several types of regions: ponds and lakes, streams and rivers, and wetlands. Ponds and lakes range in size, and small ponds may be seasonal. They sometimes have limited species diversity due to isolation from other water environments. They can get their water from precipitation, surface runoff, rivers, and springs. Streams and rivers are bodies of flowing water moving in one general direction (i.e., downstream). Streams and rivers start at their upstream headwaters, which could be springs, snowmelt or even lakes.

They continue downstream to their mouths, which may be another stream, river, lake or ocean. The environment of a stream or river may change along its length, ranging from clear, cool water near the head, to warm, sediment-rich water near the mouth. The greatest diversity of living organisms usually occurs in the middle region. Wetlands are places of still water that support aquatic plants, such as cattails, pond lilies and cypress trees. Types of wetlands include marshes, swamps and bogs. Wetlands have the highest diversity of species with many species of birds, fur-bearing mammals, amphibians and reptiles. Some wetlands, such as salt marshes, are not freshwater regions.

Marine regions cover nearly three-fourths of the earth's surface. Marine bodies are salty, having approximately 35 grams of dissolved salt per liter of water (3.5 percent). Oceans are very large marine bodies that dominate the earth's surface and hold the largest ecosystems. They contain a rich diversity of living organisms. Ocean regions can be separated into four major zones: intertidal, pelagic, benthic and abyssal. The intertidal zone is where the ocean meets the land. Sometimes, it is submerged and at other times exposed, depending upon waves and tides. The pelagic zone includes the open ocean further away from land. The benthic zone is the region below the pelagic zone, but not including the very deepest parts of the ocean. The bottom of this zone consists of sediments. The deepest parts of the ocean are known as the abyssal zone. This zone is very cold (near freezing temperatures), and under great pressure from the overlying mass of water. Mid-ocean ridges occur on the ocean floor in abyssal zones. Coral reefs are found in the warm, clear, shallow waters of tropical oceans around islands or along continental coastlines.

They are mostly formed from calcium carbonate produced by living coral. Reefs provide food and shelter for other organisms and protect shorelines from erosion. Estuaries are partially enclosed areas where fresh water and silt from streams or rivers mix with salty ocean water. They represent a transition from land to sea and from freshwater to saltwater. Estuaries are biologically very productive areas and provide homes for a wide variety of plants, birds and animals.

Deserts have relatively little vegetation and the substrate consists mostly of sand, gravel or rocks. The transition regions between deserts and grasslands are sometimes called semiarid deserts (e.g. the Great Basin of the western United States).

Grasslands cover regions where moderate rainfall is sufficient for the growth of grasses, but not enough for stands of trees. There are two main types of grasslands: tropical grasslands (savannas) and temperate grasslands. Tropical grasslands occur in warm climates such as Africa and very limited regions of Australia. They have a few scattered trees and shrubs, but their distinct rainy and dry seasons prevent the formation of tropical forests. Lower rainfall, more variable winter-through-summer temperatures and a near lack of trees characterize temperate grasslands. Prairies are temperate grasslands at fairly high elevation. They may be dominated by long or short grass species.

The vast prairies originally covering central North America, or the Great Plains, were the result of favorable climate conditions created by their high elevation and proximity to the Rocky Mountains. Because temperate grasslands are treeless, relatively flat and have rich soil, most have been replaced by farmland.

Forests are dominated by trees and can be divided into three types: tropical forests, temperate forests and boreal forests. Tropical forests are always warm and wet and are found at lower latitudes. Their annual precipitation is very high, although some regions may have distinct wet and dry seasons. Tropical forests have the highest biodiversity of this biome. Temperate forests occur at mid-latitudes (i.e., North America), and therefore have distinct seasons. Summers are warm and winters are cold. The temperate forests have suffered considerable alteration by humans, who have cleared much of the forest land for fuel, building materials and agricultural use. Boreal forests are located in higher latitudes, like Siberia, where they are known as "taiga."

They have very long, cold winters and a short summer season when most of the precipitation occurs. Boreal forests represent the largest biome on the continents.

Very low temperatures, little precipitation and low biodiversity characterize tundra. Its vegetation is very simple, with virtually no trees. The tundra can be divided into two different types: arctic tundra and alpine tundra. The arctic alpine occurs in polar regions. It has a very short summer growing season. Water collects in ponds and bogs, and the ground has a subsurface layer of permanently frozen soil known as permafrost. Alpine tundra is found at high elevations in tall mountains. The temperatures are not as low as in the arctic tundra, and it has a longer summer growing season.

Evolution of Life

Wherever they are found in the biosphere, living organisms are necessarily linked to their environment. Ecosystems are dynamic and communities change over time in response to abiotic or biotic changes in the environment. For example, the climate may be become warmer or colder, wetter or drier, or the food chain may be disrupted by the loss of a particular population or the introduction of a new one. Species must be able to adapt to these changes in order to survive. As they adapt, the organisms themselves undergo change. Evolution is the gradual change in the genetic makeup of a population of a species over time. It is important to note that it is the population that evolves, rather than individuals.

A species evolves to a particular niche either by adapting to use a niche's environment or adapting to avoid competition with another species. Recall that no two species can occupy the exact same niche in an ecosystem. The availability of resources is pivotal.

In the case of five warbler species which all consume insects of the same tree, to survive each species needs to gather its food (insects) in different parts of that tree. This avoids competition and the possible extinction of one or more species. Therefore, one of the bird species will adapt to hunting at the treetops; another the lowest branches; another the mid-section. In this way, these species have evolved into different, yet similar, niches. All five species in this way can survive by adapting to a narrow niche. Organisms with a narrow niche are called specialized species. Another example is a species that may evolve to a narrow niche by consuming only one type of leaf, such as the Giant Panda, which consumes bamboo leaves.

This strategy allows it to co-exist with another consumer by not competing with it. In both cases, species with a narrow niche are often vulnerable to extinction because they typically cannot respond to changes in the environment. Evolving to a new niche would take too much time for the specialized species under the duress of a drought, for example. On the other hand, a species that can use many foods and locations in which to hunt or gather are known as generalized species. In the event of a drought, a generalized species such as a cockroach may be more successful in finding alternative forms of food, and will survive and reproduce.

Yet another form of evolution is co-evolution, where species adapt to one another by interacting closely. This relationship can be a predator-prey type of interaction. Prey is at risk, but as a species it has evolved chemical defenses or behaviours. On the other hand, co-evolution can be a mutualistic relationship, often characterized by the ants and an acacia tree of South America. The acacia provides ants with food and a habitat, and its large projecting thorns provides protection from predators. The ants, in turn, protect the tree by attacking any animal landing on it and by clearing vegetation at its base. So closely evolved are the species that neither can exist without the other. Similar ecosystems may offer similar niches to organisms, that are adapted or evolved to that niche. Convergent evolution is the development of similar adaptations in two species occupying different yet similar ecosystems. Two species evolve independently to respond to the demands of their ecosystem, and they develop the same mechanism to do so. What emerges are adaptations that resemble look-alikes: Wings of birds and bats are similar, but evolved separately to meet the demands of flying through air. The dolphin, a mammal, shares adaptations that allow for movement through water with the extinct reptile ichthyosaur. They have similar streamlined shapes of fins, head, and nose, which make the bodies better suited for swimming. Natural selection is another process that depends on an organism's ability to survive in a changing environment. While evolution is the gradual change of the genetic makeup over time, natural selection is the force that favors a beneficial set of genes.

For example, birds migrating to an island face competition for the insects on a tropical tree. One genetic pool of a new generation may include a longer beak, which allows the bird to reach into a tropical flower for its nectar. When high populations of birds compete for insects, this ability to use the niche of collecting nectar favours that bird's survival. The long-beaked gene is passed to the next generation and the next, because birds can coexist with the insect-gathering birds by using a different niche. Through reproduction of the surviving longer-beaked birds, natural selection favours its adaptability. A species, family or larger group of organisms may eventually come to the end of its evolutionary line. This is known as extinction. While bad news for those that become extinct, it's a natural occurrence that has been taking place since the beginning of life on earth. Extinctions of species are constantly occurring at some background rate, which is normally matched by speciation. Thus, in the natural world, there is a constant turnover of species.

Occasionally large numbers of species have become extinct over a relatively short geologic time period. The largest mass extinction event in the earth's history occurred at the end of the Permian period, 245 million years ago. As many as 96 percent of all marine species were lost, while on land more than 75 percent of all vertebrate families became extinct. Although, the actual cause of that extinction is unclear, the consensus is that climate change, resulting from sea level change and increased volcanic activity, was an important factor. The most famous of all mass extinctions occurred at the boundary of the Cretaceous and Tertiary periods, 65 million years ago. About 85 percent of species became extinct, including all of the dinosaurs. Most scientists believe that the impact of a small asteroid near the Yucatan Peninsula in Mexico triggered that extinction event. The impact probably induced a dramatic change in the world climate. The most serious extinction of mammals occurred about 11,000 years ago, as the last Ice Age was ending. Over a period of just a few centuries, most of the large mammals around the world, such as the mammoth, became extinct. While climate change may have been a factor in their extinction, a new force had also emerged on the earth - modern humans. Humans, aided by new, sharp-pointed weapons and hunting techniques, may have hurried the demise of the large land mammals.

Over the years, human activity has continued to send many species to an early extinction. The best known examples are the passenger pigeon and the dodo bird, but numerous other species, many of them unknown, are killed off by over harvesting and other human-caused habitat destruction, degradation and fragmentation.

Our biosphere is the global sum of all ecosystems. It can also be called the zone of life on Earth, a closed (apart from solar and cosmic radiation) and self-regulating system. From the broadest biophysiological point of view, the biosphere is the global ecological system integrating all living beings and their relationships, including their interaction with the elements of the lithosphere, hydrosphere and atmosphere.

The biosphere is postulated to have evolved, beginning through a process of biogenesis or biopoesis, at least some 3.5 billion years ago. In a broader sense; biospheres are any closed, self-regulating systems containing ecosystems;

including artificial ones such as Biosphere 2 and BIOS-3; and, potentially, ones on other planets or moons.

Origin and Use of the Term

The term "biosphere" was coined by geologist Eduard Suess in 1875, which he defined as:

"The place on Earth's surface where life dwells."

While this concept has a geological origin, it is an indication of the impact of both Darwin and Maury on the earth sciences. The biosphere's ecological context comes from the 1920s, preceding the 1935 introduction of the term "ecosystem" by Sir Arthur Tansley. Vernadsky defined ecology as the science of the biosphere. It is an interdisciplinary concept for integrating astronomy, geophysics, meteorology, biogeography, evolution, geology, geochemistry, hydrology and, generally speaking, all life and earth sciences.

Gaia Hypothesis

The concept that the biosphere is itself a living organism, either actually or metaphorically, is known as the Gaia hypothesis.

James Lovelock, an atmospheric scientist from the United Kingdom, proposed the Gaia hypothesis to explain how biotic and abiotic factors interact in the biosphere. This hypothesis considers Earth itself a kind of living organism. Its atmosphere, geosphere, and hydrosphere are cooperating systems that yield a biosphere full of life. In the early 1970s, Lynn Margulis, a microbiologist from the United States, added to the hypothesis, specifically noting the ties between the biosphere and other Earth systems. For example, when carbon dioxide levels increase in the atmosphere, plants grow more quickly. As their growth continues, they remove more and more carbon dioxide from the atmosphere. Many scientists are now involved in new fields of study that examine interactions between biotic and abiotic factors in the biosphere, such as geobiology and geomicrobiology.

Ecosystems occur when communities and their physical environment work together as a system. The difference between this and a biosphere is simple, the biosphere is everything in general terms.

Extent of Earth's Biosphere

Every part of the planet, from the polar ice caps to the Equator, supports life of some kind. Recent advances in microbiology have demonstrated that microbes live deep beneath the Earth's terrestrial surface, and that the total mass of microbial life in so-called "uninhabitable zones" may, in biomass, exceed all animal and plant life on the surface. The actual thickness of the biosphere on earth is difficult to measure. Birds typically fly at altitudes of

650 to 1,800 meters, and fish that live deep underwater can be found down to -8,372 meters in the Puerto Rico Trench. There are more extreme examples for life on the planet: Rüppell's Vulture has been found at altitudes of 11,300 meters; Bar-headed Geese migrate at altitudes of at least 8,300 meters (over Mount Everest); Yaks live at elevations between 3,200 to 5,400 meters above sea level; mountain goats live up to 3,050 meters. Herbivorous animals at these elevations depend on lichens, grasses, and herbs.

Microscopic organisms live at such extremes that, taking them into consideration puts the thickness of the biosphere much greater. Culturable microbes have been found in the Earth's upper atmosphere as high as 41 km (25 mi) (Wainwright et al., 2003, in FEMS Microbiology Letters).

It is unlikely, however, that microbes are active at such altitudes, where temperatures and air pressure are extremely low and ultraviolet radiation very high. More likely these microbes were brought into the upper atmosphere by winds or possibly volcanic eruptions. Barophilic marine microbes have been found at more than 10 km (6 mi) depth in the Marianas Trench.

Microbes are not limited to the air, water or the Earth's surface. Culturable thermophilic microbes have been extracted from cores drilled more than 5 km (3 mi) into the Earth's crust in Sweden, from rocks between 65-75 °C. Temperature increases with increasing depth into the Earth's crust.

The speed at which the temperature increases depends on many factors, including type of crust (continental vs. oceanic), rock type, geographic location, etc. The upper known limit of microbial is 122 °C (*Methanopyrus kandleri* Strain 116), and it is likely that the limit of life in the "deep biosphere" is defined by temperature rather than absolute depth.

Our biosphere is divided into a number of biomes, inhabited by broadly similar flora and fauna. On land, biomes are separated primarily by latitude. Terrestrial biomes lying within the Arctic and Antarctic Circles are relatively barren of plant and animal life, while most of the more populous biomes lie near the equator. Terrestrial organisms in temperate and Arctic biomes have relatively small amounts of total biomass, smaller energy budgets, and display prominent adaptations to cold, including world-spanning migrations, social adaptations, homeothermy, estivation and multiple layers of insulation.

5

Environmental Conservation and Ecology

MAINSTREAMING THE ENVIRONMENT

Global Ecology, International Institutions and the Crisis of Environmental Governance Five years later, at the June 1997 Special Session of the United Nations General Assembly dedicated to the review of UNCED's implementation, the climate was rather different. Optimism had given way to disappointment and, in some cases, there was real concern about the viability of the "sustainable development" model, which relies on a framework of action that does not fully address the causes of environmental destruction. Developed countries have been unable or unwilling to stick to their promise of increasing the aid to development to 0.7% of GDP, as agreed in Rio. Countries like the United States, the largest contributor to global warming, have not shown the will to take effective action that would show a real commitment to reduce their industrial emissions. On the other hand, developing countries refused to take any further steps without the guarantee that substantive financial resources would back them or that at least the commitments taken in Rio would be respected. The New York 1997 Declaration even recognized that the situation of the environment had deteriorated over the intervening five years, hoping modestly that more progress would be achieved by the next summit in 2002.

The meager positive results produced by the massive efforts in the field of international cooperation for the environment seem to indicate the contradictory character of this new, global "environmentalism." The purpose of this article is to demonstrate that, while originally being the potential source of a radical and transformative project, environmental concerns were ultimately reframed by the joint action of technocratic environmentalists, the international UN-related establishment and business and industry sectors to become compatible with global development. Adopting an international political economy perspective, the article explores the interaction between state and markets in the construction of global environmental politics. It provides evidence that although there is a new consensus on the diagnosis of the problem - worldwide environmental degradation - very few commitments have been taken to alter the accumulation model and the patterns of production and consumption that contribute to this situation.

It suggests that the failure of the international system in ensuring a move towards sustainability, exemplified in New York, is linked to the very nature of the global bargain struck in Rio. By aiming to make "development" - in its more recent global phase, with its focus on globalized and ever expanding production, trade and consumption - become "sustainable," the concept of sustainability has been stripped of most of its meaning. The inability of the international community to deal with most global environmental issues reveals the contradictory nature of the "sustainable development" consensus and demonstrates the limits of international cooperation in the name of the environment.

Origins and Dimensions of the Ecological Project

In order to understand the meaning of the transformation of environmental concerns into a widely accepted concept, it is useful to recall the original purpose of the ecological project. The ecological movement finds its origins in a protest aimed at defending the right of individuals to regain influence over their ways of living, of producing, and of consuming. As stressed by Gorz (1992), it started as a radical cultural movement, as an attempt by individuals to control and understand the consequences of their actions. With the ecological critique, activists hoped to refocus attention on local knowledge and practices and to bridge the separation of humans from nature, a division that had been at the heart of the Enlightenment project.

In the 1970s, the ecological movement became a political movement, and there was an awareness that the demands of ecology were not only sectorial and local aspirations but rather represented a value shared across national divides (Smith 1996; Gorz 1992).

The publication of the report "Limits to Growth" by the Club of Rome in 1972 gave a scientific backing to these cultural demands and showed the risks posed by the model of industrial growth on the future of life on earth. The report provided a holistic view of the interrelationship between population growth, food production and consumption, the industrialization process, depletion of nonrenewable resources and waste and pollution at the global level, recognizing that waste and pollution are not only a problem for the living conditions and consumption patterns of the population, but affect the very basis of the productive sphere's reproduction. For the first time, environmental degradation provoked by economic growth was considered from a global perspective, going beyond the occasional questioning of pollution problems during the 1950s and 1960s.

In addition, the report launched a real debate on the morality of growth and of the differences in consumption and living standards between developed and developing countries. The 1970s also represented an inflection in the history of social mobilization and collective action with the emergence of the "new social movements," which identify themselves as value movements carrying universal interests going beyond class, nation, sex and race borders. The new social movements such as the environmental movement appear as "modern" in the sense that they are based upon the belief that history's course can be changed by social actors and are not determined by what Touraine calls a "metasocial principle" (Offe 1988, 219). Environmentalists believe that, although representing a real challenge to our present lifestyles and habits, it is possible to move towards a sustainable society that respects nature and privileges well-being over accumulation. Speaking about the existence of a unique and unified "green movement" is clearly incorrect. Environmental concerns mean different things to different people, take many forms and are expressed through different channels. In addition, environmentalism takes very different forms in developed or in developing countries. It can mean fighting for an even better quality of life in advanced countries, and fighting for subsistence or even survival in poor countries. Despite this diversity, for the purpose of academic inquiry, three main components of the "green movement," albeit sometimes overlapping, can be distinguished. These three categories should be viewed as "ideal-typical" and not necessarily mutually exclusive.

The first tendency of the ecological movement, deep ecology, is typically a postmodern movement.

In philosophical terms, deep ecology challenges the separation between humans and nature that was at the heart of modern humanism. Deep ecology is not "anthropocentric," it is "ecocentric." As observed by Merchant (1992), it seeks a total transformation in science and in worldviews that will lead to the replacement of the mechanistic paradigm (which has dominated the past three hundred years) by an ecological framework of interconnectedness and reciprocity. The ideas of deep ecology have influenced (among others) Greenpeace, the largest green NGO, which claims that humanist value systems must be replaced by supra-humanist values that place any vegetal or animal life in the sphere of legal and moral consideration. Greenpeace is therefore an example of an environmental organization which, based on scientific reports and examinations, acts to change worldviews and consciousness in order to promote a shift to "ecocentrism" rather than trying to act to transform the production systems which lie at the root of environmental problems.

Yet, while having influenced the most well-known environmental NGO, deep ecology remains a fairly marginal wing of the green movement. Deep ecologists have been criticized for their lack of a political critique, failing to recognize that the idea itself of "ecocentrism" is "anthropocentric." As stressed by Merchant, deep ecologists take the character of capitalist democracy for granted rather than submitting it to a critique. Their tendency to refuse to consider economic policy and to assume a purely conservationist standpoint relegates them to a secondary position.

The second component of the "green movement" is what can be called the "social ecology" movement, which is to a large extent composed of people from the "New Left," dissatisfied with Marxism. Contrary to the deep ecologists, social ecologists maintain an anthropocentric perspective: the concern for nature is understood as a concern for the environment of human beings. Social ecologists seek transformations in production and reproduction systems, that is, a transformation of political economy, as the way to achieve sustainability, social equity and well being. Social ecologists see a contradiction between the logic of capitalism and the logic of environmental protection. For them, environmental protection cannot be made dependent upon economic development, because development, in its liberal sense, has meant the subordination of every aspect of social life to the market economy, and can therefore no longer be considered as a desirable goal. The hegemonic view on "sustainable development," which rehabilitates development as the global goal of humans, is thus unsatisfactory. Social ecologists call for a rethinking of the theoretical basis of development that should include not only economic but also political and epistemological dimensions, such as the questions of participation, of empowerment and local knowledge systems.

For them, what makes development "unsustainable" at the global level is the pattern of consumption in rich countries. Thinking about sustainability thus implies considering the contradictions imposed by the structural inequalities of the global system. Finally, social ecologists vary to a certain extent in the North and in the South: generally speaking, organizations in the North sometimes carry their rejection of development as far as to strike postmodern stances, while organizations in the South focus more on equity and on the need to redistribute the benefits of development.

Finally, there is a more technocratic tendency to the green movement, a tendency that tries to make economic growth and environmental protection appear as compatible goals, which need not require a profound change in values, motivations and economic interests of social actors, nor new models of economic accumulation. For them, it is because capitalist production methods and life standards are not developed enough that environmental problems emerge.

The evidence is that environmental standards are higher in richer countries. Technocratic environmentalists seek to preserve the environment through the establishment of international institutions, the use of economic and market instruments and the development of clean and "green" technology. The result is a rather apolitical approach and activists who, though still interested in environmental protection, are not primarily committed to ideas of equity and social justice, or at least not as committed as social ecologists (Gudynas 1993). The technocratic tendency is thus essentially a rich country tendency, although it is also present in some elite circles in the South.

These environmentalists tend to focus on issues of population for example, arguing that the biggest threat to the environment comes from high population growth in the Third World and the pressure it will bring to bear on the stock of natural resources. Technocratic environmentalists usually tend to belong to organizations which have little or no membership, and rely on their technical and legal expertise and on their research and publishing programs to influence decision-making. Through their close relationship with government and other influential actors and their easy access to international organizations, these organizations tend to have a greater impact than activist membership organizations.

Today, it can be said that this technocratic approach appears to be prevailing over both the biocentric (deep ecology) and the social ecology perspectives and has become what is today mainstream environmentalism, which finds its major expression in the concept of "sustainable development." Despite the challenging and radical nature of ecological concerns, the fact that they might present a potential for change in the present economic model, they were ultimately reframed so as to constitute what appears as an apolitical, techno-managerial approach.

The Formation of a Consensus on "Sustainable Development" It is interesting to examine how the apparent consensus around the concept of "sustainable development" was built and how the project of global environmental "management" became hegemonic. Two main actors have contributed to the hegemony of the liberal environmental management project.

One is the scientific and policy-making environmental community, or, in the words of Peter Haas, the environmental "epistemic community" (Haas 1990); the other actor is business and industry.

The Brundtland Report, the United Nations Conference and the Global North-South "Bargain" International environmental politics did not emerge in the 1990s. As early as 1972, a United Nations Conference on the Human

Environment took place in Stockholm, launching the era of international environmental negotiations.

Stockholm did produce some significant outcomes, leading to the creation of the United Nations Environment Program (UNEP), based in Nairobi, which coordinates environmental action within the United Nations. The context of the Stockholm Conference was not very favorable to the adoption of strong environmental commitments. Developing countries were unsatisfied with the UN system and preparing the movement for a New International Economic Order.

They were not willing to yield part of their sovereignty over natural resources in the name of environmental protection, and denounced the emergence of "eco-imperialism." The oil crisis of the 1970s relegated environmental protection to a marginal position in international relations. In the 1980s, the international climate started to change as the debt crisis was seriously affecting developing countries and their role and participation in international fora. In this context, "international commissions" were established to try to elaborate global proposals to promote peace and development, such as the Brandt Commission. Efforts were also undertaken to replace environmental protection on the international political agenda. The World Commission on Environment and Development was established in 1983 under the presidency of Gro Harlem Brundtland, and asked to produce a comprehensive report on the situation of the environment at the global level.

The work of the Commission represented a landmark in international initiatives to promote environmental protection as it produced the concept of sustainable development, a concept that would become the basis of environmental politics worldwide. Sustainable development is defined by the Brundtland Report as a development that is "consistent with future as well as present needs" (World Commission on Environment and Development 1987).

The concept of sustainable development was built as a political expression of the recognition of the "finiteness" of natural resources and of its potential impact on economic activities. Indeed, the report argues that, while we have in the past been concerned about the impacts of economic growth upon the environment, we are now forced to concern ourselves with the impacts of ecological stress - degradation of soils, water regimes, atmosphere and forests- upon our economic prospects. The report offered a holistic, global vision of today's situation by arguing that the environmental crisis, the developmental crisis and the energetic crisis are all part of the same, global crisis. It offers solutions to this global crisis, which are mainly of two kinds. On the one hand there are solutions based on international cooperation, with the aim of achieving an international economic system committed to growth and the elimination of poverty in the world, able to manage common goods and to provide peace, security, development and environmental protection. On the other hand, come recommendations aiming at institutional and legal change, including measures not only at the domestic level but also at the level of international institutions. The report emphasizes the expansion and improvement of the growth-oriented industrial model of development as the way to solve the global crisis.

The Brundtland Report also promoted the view that global environmental degradation can be seen as a source of economic disruption and political tension, therefore entering the sphere of strategic considerations. For the Brundtland Commission, the traditional forms of national sovereignty are increasingly challenged by the realities of ecological and economic interdependence, especially in the case of shared ecosystems and of "global commons," those parts of the planet that fall outside national jurisdictions. Here, sustainable development can be secured only through international cooperation and agreed regimes for surveillance, development, and management on the common interest.

For example, the consequences of climate change such as rising sea levels and the effects of temperature variations on agricultural production would require deep changes in the economy and impose high costs on all countries, thus leading to very unstable situations. The issue of forest preservation can also fit into this context, since forests contribute to the stability of climate by acting as carbon sinks, and assure the regeneration of ecosystems by providing reservoirs of biological diversity. Preserving forests then becomes more than an ecological concern: it is also a security imperative. So the "environmental security" discourse was also a cause for the need to find a "consensual solution" to issues of environmental protection. The United Nations Conference on Environment and Development (UNCED), held in Rio de Janeiro in June 1992, marked the official institutionalization of environmental issues in the international political agenda.

Twenty years after the 1972 Stockholm Conference, which was on the "Human Environment," Rio meant a real shift in the vision that had dominated environmental politics so far. After Rio, environmental considerations became incorporated into development, and a "global bargain" was struck between North and South on the basis of the acceptance from both sides of the desirability of achieving a truly global economy which would guarantee growth and better environmental records to all. UNCED recognized the "global finiteness" of the world, i.e., the scarcity of natural resources available for development, but adopted the view that, if the planet is to be saved, it will be through more and better development, through environmental management and "eco-efficiency." The UNCED process involved over a hundred and fifty hours of official negotiations spread over two and a half years, including two planning meetings, four Preparatory Committees (Prepcoms), and the final negotiation session at the Rio Summit in June 1992.

The major result of UNCED is called "Agenda 21," a 700-page global plan of action which should guide countries towards sustainability through the 21st century, encompassing virtually every sector affecting environment and development. Besides Agenda 21, UNCED produced two non-binding documents, the "Rio Declaration" and the Forest Principles. In addition, the climate change and the biodiversity conventions, which were negotiated independently of the UNCED process in different fora, were opened for signature during the Rio Summit and are considered as UNCED-related agreements. The "Rio Declaration," which was the subject of much dispute between the Group of 77 (the coalition of developing countries) and industrialized countries, mainly the United States, illustrates well the kind of bargain reached in Rio.

It recognizes the "right of all nations to development" and their sovereignty over their national resources, identifies "common but differentiated responsibility" for the global environment, and emphasizes the need to eradicate poverty, all demands put forward by the Group of 77. In return, the suggestions by the G77 to include consumption patterns in developed countries as the "main cause" of environmental degradation and the call for "new and additional resources and technology transfer on preferential and concessional terms" were rejected by OECD countries.

In the end, on the issue of finance, an institution called the "Global Environment Facility" (GEF) was set, under the joint administration of the World Bank, the United Nations Development Program (UNDP) and the United Nations Environment Program (UNEP), as the only funding mechanism on global environmental issues, and OECD countries committed themselves to achieving a target of 0.7 percent of GNP going to ODA (Overseas Development Assistance) by the year 2000, to help developing countries implement UNCED's decisions.

Despite the failure of the G77 to win significant concessions on financial resources, if one considers the differences in priorities between developed and developing countries and the conflictual character of the negotiation process, UNCED's outcomes were still seen by the international establishment as quite impressive, marking "an important new stage in the longer-term development of national and international norms and institutions needed to meet the challenge of environmentally sustainable development."

A Commission on Sustainable Development (CSD) was established to monitor and report on progress towards implementing UNCED's decisions. In particular, the CSD's stated aims are to enhance international cooperation by rationalizing the intergovernmental decision-making capacity, and to examine progress in the implementation of Agenda 21 at the national, regional and international levels.

After UNCED, environmental considerations were "integrated" at all levels of action. The "sustainable development paradigm," as some authors recognize, is already replacing the "exclusionist paradigm" (i.e., the idea of an infinite supply of natural resources) in some multilateral financial institutions, as well as in some state bureaucracies and in some parliamentary committees. Most economists now acknowledge that natural resources are scarce and have a value that should be internalized in costs and prices.

Organizations such as the European Union made the "integration" of environmental concerns one of their leading policy principles.

Many countries carried out environmental policy reform to implement UNCED's decisions and the Agenda 21. The boundaries of environmental politics were broadened and its links with all other major issues on the international arena, such as trade, investments, debt, transports, for example, were examined. Efforts were also undertaken to improve environmental records of multilateral finance and development institutions. The World Bank, which has a long history of contributing to environmental degradation by financing destructive projects, went through a "greening" process, and now has a "Department of the Environment" which conducts "environmental impact assessments" and imposes "environmental conditionalities" before granting loans. The World Trade Organization has a "Committee on Trade and Environment" (CTE) which is in charge of ensuring that open trade and environmental protection are mutually supportive. All these efforts can be seen, according to Porter and Brown (1996), as part of a longer-term process of evolution toward environmentally sound norms governing trade, finance, management of global commons, and even domestic development patterns. Environmental considerations were then to be introduced in all major international bureaucracies as a dimension to take into consideration in decision-making processes, and as a challenge for global management. To a certain extent, the "technocratic" approach became hegemonic because it best suited the interests of the international development elite as it magnified its managerial responsibilities. In a time when the legitimacy and utility of the United Nations system was being seriously questioned by its idealizer and major financial supporter - the United States - the goal of making environment and development compatible was seized by some UN agencies as an unexpected opportunity to regain credibility, as well as to be granted funds and to hire new staff for recently created units on "trade and environment" or "finance and environment." UNCED provided a new legitimacy to international organizations such as the World Bank or the World Trade Organization and to their bureaucracies, which now try to assume a leading role in "managing the earth." With the promotion of economic growth to a planetary imperative and the rehabilitation of technological progress, both development institutions and organizations and states appeared as legitimate agents to solve global environmental problems.

If international organizations have benefited from the global perspective that emerged from Rio, they have also contributed to mold it. There is an active "epistemic community," which includes both the international organization establishment and large environmental NGOs, promoting the "global environmental management" approach.

These groups tend to believe that their moral views are cosmopolitan and universal, and emphasize the existence of an international society of human beings sharing common moral bonds.

In this kind of "same boat" ideology, environmental concerns tend to be presented as moral imperatives, related neither to political nor to economic advantages. It would be a consensual concern, a sort of universal principle accepted over borders and political boundaries. An example of an institution promoting these ideas is given by the Commission on Global Governance. In the words of the Commission, "we believe that a global civic ethic to guide action within the global neighborhood and leadership infused with that ethic are vital to the quality of global governance. We call for a common commitment to core values that all humanity could uphold. We further believe humanity as a whole will be best served by recognition of a set of common rights and responsibilities."

Part of the Green movement came to support this "same boat ideology" and was incorporated into the epistemic community. Actually, mainstream conservationist environmentalists were fully admitted into the global environmental management establishment, conferring legitimacy to the UNCED process.

NGOs contributed to UNCED to a degree unprecedented in the history of UN negotiations. NGOs lobbied at the official process, participated in Prepcoms and were even admitted in some countries' delegations, a novelty which was rendered possible by resolution 44/228 calling for "relevant non-governmental organizations in consultative status with the Economic and Social Council to contribute to the Conference, as appropriate."

In addition, during UNCED, NGOs organized in Rio a meeting which ran parallel to the official governmental conference. The "Global Forum," which gathered about 30,000 people, represented 760 associations, among participants and visitors, in a sort of "NGO city." During one week, the Global Forum became home to environmentalists and social activists, to Indians and ethnic minorities, and to feminists and homosexual groups, all united to "save the earth." NGOs organized many demonstrations protesting against the modest results of the official summit and elaborated their own agenda for improving environmental protection worldwide.

Yet, in the eyes of some observers, NGO efforts tended to become coopted by larger and richer groups from advanced countries, which had more means, not only financially but also in terms of organizational, scientific and research capac- ity, to promote their own views (Chatterjee and Finger 1994).

In the end, NGOs decided that they would sign, in Rio, NGOs "treaties" on all the issues being discussed at the UNCED official meeting. The main activity at the Global Forum was then the "treaty negotiation" process, just like at the official forum, a process which proved to be very disappointing, as the same North-South conflicts that were blocking UNCED tended to separate northern and southern NGOs.

Ultimately, the NGO treaty process was little more than a pantomime of real diplomacy, and ultimately, the treaties agreed upon, negotiated among a couple of dozen NGOs, had a very modest impact on the future of NGO activities.

The representation at the Global Forum was also very unequal, illustrating differences in means between northern NGOs, very present, and southern NGOs. Asian, and above all, African NGOs, were severely under-represented. Differences in associative traditions and language barriers also explain the hegemony of Anglo-Saxon organizations at the Global Forum. In the end, influential NGOs decided to concentrate their efforts on lobbying the official conference. The Earth Summit in 1992 thus represented a real moment of acceleration for NGO activities, as it allowed some of them to have a better idea of what their counterparts were doing in other parts of the world, and was the base for establishing cooperation projects and partnerships among organizations. Yet while NGO efforts illustrated by the Global Forum aimed at uniting NGOs worldwide, the green movement came out of Rio appearing even weaker and more fragmented, with the polarization between "realist," co-operative NGOs on the one side and "radical," transformative NGOs on the other.

Finally, the "sustainable development" approach also suited the interests of some governments in the Third World which are primarily committed to economic development and sought through UNCED to obtain concessions in financial and technological terms in exchange of their support for environmental management. Some Third World countries are still marked by a "developmentalist" ideology in which economic development comes before all else.

In addition, resource rich countries such as Malaysia, Indonesia, or Brazil, have traditionally had a vision of unending and expanding frontiers, in which land and natural resources are unlimited and no constraints are seen to exist on the use of resources. As a result, they were unwilling to accept the elaboration of international regimes aiming at limiting their sovereignty over the exploitation of natural resources.

The issue of sovereignty had long been a major source of tension during international environmental negotiations. As long ago as the Stockholm Conference in 1972 developing countries had pressed for the inclusion of a specific principle on the topic. Principle 21 of the Stockholm Declaration stated that "States have, in accordance with the Charter of the United Nations and the principles of international law, the sovereign right to exploit their own resources pursuant to their own environmental policies, and the responsibility to ensure that activities within their jurisdiction or control do not cause damage to the environment of other States or areas beyond the limits of national jurisdiction."

The same debate arose when UNCED was convened, and in the end the sovereignty principle as in stood in the Stockholm Declaration's Principle 21 was included in the Rio Declaration.

In addition, a guarantee that economic development would continue to be the priority on the international agenda was an essential element for developing countries. The reaffirmation of the right to development, and of the sovereignty principle, ensured in Rio, were then the two elements that made agreement at UNCED possible for the Group of 77. The alliance between environment and development could then become official. As described by the vice-president of the International Institute for Environment and Development (IIED), "it has not been too difficult to push the environment lobby of the North and the development lobby of the South together. And there is now in fact a blurring of the distinction between the two, so they are coming to have a common consensus around the theme of Sustainable Development" (World Commission on Environment and Development 1987, 64). Yet to fully understand the nature of this consensus around sustainable development, one last actor needs to be introduced. The actor whose vision shaped most fundamentally the content of this consensus and the real winner of Rio, the business and industry sector, and in particular transnational corporations.

ECOLOGICAL SECURITY

Spiraling population and increasing industrialisation are posing a serious challenge to the preservation of the terrestrial and aquatic ecosystems. Environmental protection is the key to ensuring a healthy life for the people. Environmental problems are on the increase and are more pronounced in densely populated cities. Creation of awareness regarding the ecological hazards among the public is absolutely essential. Environmental conservation and abatement of pollution are critical for sustainable development.

Department of Environment

The Department of Environment was created in 1995 as the nodal Department for dealing with environmental management of the State.

The Department is entrusted with the implementation of major projects like pollution abatement in the Cauvery, Vaigai and Tamiraparani rivers, pollution abatement in Chennai city waterways, National Lake Conservation Programme and all aspects of environment other than those dealt with by Tamil Nadu Pollution Control Board.

One of the main objectives of the department is to implement Environmental Awareness Programme Wide publicity is being given on World Environment Day, Ozone day and on Bhogi Day to create environmental awareness among the general public. Due to the concerted efforts, the level of air and noise pollution has been brought down to the tune of 20-25% in the last three years. To create environmental awareness among the school and college studies, 1260 eco-clubs have been formed in all the districts of State involving selected educational institutions and NGOs. It is proposed to strengthen these existing eco-clubs. The outstanding NGOs, experts and individuals are honoured with environmental awards in recognition of their excellent contribution in the field of environment.

Tamil Nadu Pollution Control Board

The Tamil Nadu Pollution Control Board enforces the provisions of the Water (Prevention and Control of Pollution) Act, 1974 as amended, the Water (Prevention and Control of Pollution) Cess Act, 1977 as amended, the Air (Prevention and Control of Pollution) Act, 1981 as amended and the relevant provisions/rules of the Environment (Protection) Act, 1986 to prevent, control and abate pollution and for protection of environment.

The Board functions with its Head Office at Chennai. There are 25 District Offices and 14 Environmental Laboratories established by the Board.

Monitoring of Industries

The Board has inventorised about 28,000 industries. The Board has prescribed standards for discharge of effluent, ambient air quality and gaseous emissions from various industries and the industries have to take necessary pollution control measures to meet the standards prescribed by the Board. For effective monitoring, the Board has classified the industries into red, orange and green, based on their pollution potential.

Procedure for Issue of Consent

The Board issues consent to industries in two stages under the Water Act and the Air Act for establishment and operation of industrial units. Consent to establish is issued depending upon the suitability of the site, before the industry takes up the construction activity. Consent to operate is issued after installation of effluent treatment plant and air pollution control measures, before commissioning production. Consent is issued subject to general conditions and specific conditions.

Common Effluent Treatment Plants

The Board plays an important role in the establishment of Common Effluent Treatment Plants (CETPs) for clusters of small polluting industries in various parts of the State. Tamil Nadu is a pioneering State in India in establishing CETPs. So far, proposals for 50 CETPs have been formulated. Of these, 33 CETPs are under operation and 17 CETPs are under various stages of implementation. Towards the implementation of CETPs, State subsidy is granted by Government of Tamil Nadu, upto 25% of the project cost and Central subsidy is granted by Government of India, upto 25% of the project cost. The details of CETPs are as follows:

Sl. No.	Sector	No. of CETPs formed	No. of CETPs under operation
1	Tanneries	24 schemes	14 schemes
2	Textile Bleaching and Dyeing	25 schemes	18 schemes
3	Hotels and Lodging Houses	1 scheme	1 scheme
	Total	50 schemes	33 schemes

Cleaner Technologies & Water and Energy Conservation

With active support and encouragement from the Board, industrial units in Tamil Nadu have started switching over to cleaner technologies and also started water and energy conservation measures.

Air quality monitoring As per the provisions of the Air Act, the entire State of Tamil Nadu has been declared as air pollution control area. The Board is monitoring the ambient air quality in Chennai (3 stations), Coimbatore (3 stations), Thoothukudi (3 stations), Madurai (3 stations) and Salem (1 station) under the National Ambient Air Quality Monitoring Programme. Under the State Ambient Air Quality Monitoring Programme, the Board has established 5 ambient air quality monitoring stations in Chennai city and 5 in Tiruchirapalli. Towards preparation of the environmental management plan for Chennai city, ambient air quality surveys have been conducted at 41 stations in Chennai to identify the most sensitive locations with respect to air pollution. The Board has established 6 continuous ambient air quality monitoring systems at Cuddalore, Thoothukudi, Ranipet, Manali-Thiruvallur, Royapuram-Chennai, Kathivakkam-Thiruvallur to assess the level of pollutants in the ambient air and the adequacy of air pollution control measures provided by the industries in the vicinity.

Vehicle Emission Monitoring

The Board is carrying out the vehicle emission monitoring in Chennai, Dindigul, Palani, Udhagamandalam and Chengalpattu. In addition, private agencies have been authorised by the Transport Department in Chennai city to check the emission level of the vehicles.

The Board has upgraded and computerised all its vehicle emission monitoring stations for testing diesel driven vehicles. The Transport Corporations have also been instructed to closely monitor the emission levels of their buses.

For controlling vehicular emission, cleaner fuel like unleaded petrol, petrol with 3% benzene and low sulphur fuel (0.05%) have been introduced in Chennai Metropolitan Area. Passenger cars complying with Bharat stage-II norms alone are registered in Chennai since July 2001. 2T oil auto dispensing system have been provided in retail outlets.

The Board is also participating in a research project with an non governmental organisation and the Civil Supplies Department to study the use of gas chromatograph to detect fuel adulteration. Action has already been taken to introduce auto liquefied petroleum gas in Chennai as it is a cleaner fuel. Steps are being taken to popularise the use of liquefied petroleum gas for autorickshaws, call taxis and other private vehicles which will help in improving air quality.

(1) Vehicle Emission Monitoring Stations provided by the Board. 8 Nos.
(2) Vehicle Emission Monitoring Stations provided by the Board in MTC Depots. 6 Nos.
(3) Emission Checking Stations provided by Private Agencies. 236 Nos.
(4) Auto Liquefied Petroleum Gas Dispensing Stations commissioned in Chennai. 12 Nos.

Noise Level Monitoring

Towards controlling noise pollution in urban areas, about 52,586 air horns were removed as of December 2004 from buses and lorries throughout the State. All the districts have been declared as air horn free districts. For noise level monitoring at the district level, sophisticated noise level meters have been provided to the District Offices of the Board.

Water Quality Monitoring

Pollution of major rivers in the State is caused by the discharge of untreated sewage from the urban local bodies and panchayats and untreated or partially treated effluent from industries. In case of industrial pollution, it is the responsibility of the industrial units to provide the required effluent treatment plants either individually or collectively so as to achieve the standards. Various pollution abatement schemes are being implemented under the National River Conservation Programme under the coordination of the Department of Environment.

Water Quality Monitoring Programmes

Under the Global Environmental Monitoring System, the Board is closely monitoring the quality of water in the Cauvery basin at Mettur, Pallipalayam, Musiri and ground water quality at Musiri. Similarly, water quality of rivers Cauvery (16 stations), Tamiraparani (7 stations), Palar (1 station) and Vaigai (1 station) and the three important lakes in Udhagamandalam, Kodaikkanal and Yercaud are being monitored under the Monitoring of Indian National Aquatic Resources System by the Board. The Board is continuously monitoring the Chennai city water ways to prevent pollution due to discharge of trade effluent from industries and sewage from local bodies and is collecting and analysing samples of river water and outfalls at regular intervals, since 1991.

ECOLOGY AND ENVIRONMENT OF MANGROVE ECOSYSTEMS

The mangrove environment has some special physicochemical characteristics of salinity, tidal currents, winds, high temperatures, and muddy anaerobic soil.

Plants of the environment are able to adapt to themselves to practically all types of adverse conditions except perhaps frost, and hence they are distributed mostly in the tropical regions. Probably there are no other groups of plants with such highly developed adaptations to extreme conditions. Mangroves occur in low-lying, broad coastal plains where the topographic gradients are small and the tidal amplitude large. Repeatedly getting flooded, but well-drained soil supports a rich growth of mangrove plants.

They normally grow poorly in stagnant waters and have luxuriant growth in the alluvial soil substrates with fine-textured loose mud or silt, rich in humus and sulphides. They can also be found in substrates other than muddy soil such as coastal reefs and oceanic islands. In such areas, the mangrove plants grow on peat, which is derived from decayed vegetation. They find it difficult to colonize the coastal zone with waves of high energy and hence they normally establish themselves in sheltered shorelines (Kathiresan & Bingham, 2001).

Types of Coastal Settings

Mangroves get tightly bound to the coastal environments in which they occur. Not only are they influenced by chemical and physical conditions of their environment, but they usually help to create those conditions by themselves. They are found in a variety of tropical coastal settings like the deltas, estuarine areas with their own deltas, lagoons, and fringes of the coral reefs.

There are normally six functional types of mangrove forests, namely, fringe, riverine, basin, overwash, scrub (dwarf) and hommock forests. The last three types are the modified forms of the first three types. The six types can be summarized as follows:

1. *Overwash Mangrove Forests:* These are small mangrove islands, frequently formed by tidal washings.
2. *Fringing Mangrove Forests:* These occur along the borders of protected shorelines and islands, influenced by daily tidal range. They are sensitive to erosion and long exposure to purely marine conditions with turbulent waves, and tides.
3. *Riverine Mangrove Forests:* These are luxuriant patches of mangroves existing along rivers and creeks, which get flooded daily by the tides. Such forests are influenced with the incursion of large amount of freshwater with fluvial nutrients and thus making the system highly productive with trees growing taller.
4. *Basin Mangrove Forests:* These are stunted mangroves located along the interior side of the swamps and in drainage depressions. Their positioning channels the terrestrial runoff to move towards the coast with a slow velocity of water flow.
5. *Hammock Mangrove Forests:* These are similar to the basin type except that they occur in more elevated sites than the four types given above.
6. *Scrub Mangrove Forests:* These form dwarf mangrove settings along flat coastal fringes.

Based on the substrate, tidal range and sedimentation, six more broad classes of mangrove settings have been given as follows by Thom (1982) and Galloway (1982):

1. Large deltaic systems (occurring in low tidal range, very fine allochtonous sediments) (e.g. mangroves of Borneo, Sundarbans)
2. Tidal plains (where alluvial sediments are reworked by the tides; and there is the presence of large mudflats for the growth of mangroves)
3. Composite plains; under the influence of both tidal and alluvial conditions (e.g. lagoons formed behind wave-built barriers where mangroves grow)

4. Fringing barriers with lagoons (high wave energy conditions with autochtonous sediments of fine sand and mud) (e.g. mangroves of the Philippines),
5. Drowned bedrock valleys (e.g. mangroves of Northern Vietnam or Eastern Malaysia)
6. Coral coasts (mangroves growing at the bottom of coral sand or in platform reefs) (e.g. mangroves of India, Indonesia and Singapore)

Most of the types of mangrove forests noted above get frequently affected by rapid changes in coastal geomorphology. Many ecological factors strongly influence the well-being of mangroves and these include geographical latitudes, wave action, rainfall, freshwater runoff, erosion/sedimentation rates, aridity, salinity, nutrient inputs, and soil quality. Primary productivity and biomass of mangroves normally decrease with increasing latitudes.

ECOLOGICAL FACTORS: DYNAMICS AND STABILITY

Ecological factors which affect dynamic change in a population or species in a given ecology or environment are usually divided into two groups: abiotic and biotic. Abiotic factors are geological, geographical, hydrological and climatological parameters. A biotope is an environmentally uniform region characterized by a particular set of abiotic ecological factors. Specific abiotic factors include:

- Water, which is at the same time an essential element to life and a milieu
- Air, which provides oxygen, nitrogen, and carbon dioxide to living species and allows the dissemination of pollen and spores
- Soil, at the same time source of nutriment and physical support
- Soil pH, salinity, nitrogen and phosphorus content, ability to retain water, and density are all influential
- Temperature, which should not exceed certain extremes, even if tolerance to heat is significant for some species
- Light, which provides energy to the ecosystem through photosynthesis
- Natural disasters can also be considered abiotic

Biocenose, or community, is a group of populations of plants, animals, micro-organisms. Each population is the result of procreations between individuals of same species and cohabitation in a given place and for a given time. When a population consists of an insufficient number of individuals, that population is threatened with extinction; the extinction of a species can approach when all biocenoses composed of individuals of the species are in decline. In small populations, consanguinity (inbreeding) can result in reduced genetic diversity that can further weaken the biocenose. Biotic ecological factors also influence biocenose viability; these factors are considered as either intraspecific and interspecific relations. Intraspecific relations are those which are established between individuals of the same species, forming a population. They are relations of co-operation or competition, with division of the territory, and sometimes organization in hierarchical societies.

An antlion lies in wait under its pit trap, built in dry dust under a building, awaiting unwary insects that fall in. Many pest insects are partly or wholly controlled by other insect predators.

Interspecific relations-interactions between different species-are numerous, and usually described according to their beneficial, detrimental or neutral effect (for example, mutualism (relation ++) or competition (relation --). The most significant relation is the relation of predation (to eat or to be eaten), which leads to the essential concepts in ecology of food chains (for example, the grass is consumed by the herbivore, itself consumed by a carnivore, itself consumed by a carnivore of larger size). A high predator to prey ratio can have a negative influence on both the predator and prey biocenoses in that low availability of food and high death rate prior to sexual maturity can decrease (or prevent the increase of) populations of each, respectively. Selective hunting of species by humans which leads to population decline is one example of a high predator to prey ratio in action. Other interspecific relations include parasitism, infectious disease and competition for limiting resources, which can occur when two species share the same ecological niche.

The existing interactions between the various living beings go along with a permanent mixing of mineral and organic substances, absorbed by organisms for their growth, their maintenance and their reproduction, to be finally rejected as waste. These permanent recyclings of the elements (in particular carbon, oxygen and nitrogen) as well as the water are called biogeochemical cycles.

They guarantee a durable stability of the biosphere (at least when unchecked human influence and extreme weather or geological phenomena are left aside).

This self-regulation, supported by negative feedback controls, ensures the perenniality of the ecosystems. It is shown by the very stable concentrations of most elements of each compartment. This is referred to as homeostasis. The ecosystem also tends to evolve to a state of ideal balance, reached after a succession of events, the climax (for example a pond can become a peat bog).

SPATIAL RELATIONSHIPS AND SUBDIVISIONS OF LAND

Ecosystems are not isolated from each other, but are interrelated. For example, water may circulate between ecosystems by the means of a river or ocean current. Water itself, as a liquid medium, even defines ecosystems. Some species, such as salmon or freshwater eels move between marine systems and fresh-water systems. These relationships between the ecosystems lead to the concept of a biome. A biome is a homogeneous ecological formation that exists over a large region as tundra or steppes. The biosphere comprises all of the Earth's biomes -- the entirety of places where life is possible -- from the highest mountains to the depths of the oceans.

Biomes correspond rather well to subdivisions distributed along the latitudes, from the equator towards the poles, with differences based on to the physical environment (for example, oceans or mountain ranges) and to the climate. Their variation is generally related to the distribution of species according to their ability to tolerate temperature and/or dryness. For example, one may find photosynthetic algae only in the photic part of the ocean (where light penetrates), while conifers are mostly found in mountains.

Though this is a simplification of more complicated scheme, latitude and altitude approximate a good representation of the distribution of biodiversity within the biosphere. Very generally, the richness of biodiversity (as well for animal than plant species) is decreasing most rapidly near the equator and less rapidly as one approaches the poles. The biosphere may also be divided into ecozones, which are very well defined today and primarily follow the continental borders. The ecozones are themselves divided into ecoregions, though there is not agreement on their limits.

Ecosystem Productivity

In an ecosystem, the connections between species are generally related to food and their role in the food chain. There are three categories of organisms:

- Producers -- usually plants which are capable of photosynthesis but could be other organisms such as bacteria around ocean vents that are capable of chemosynthesis.
- Consumers -- animals, which can be primary consumers (herbivorous), or secondary or tertiary consumers (carnivorous and omnivores).
- Decomposers -- bacteria, mushrooms which degrade organic matter of all categories, and restore minerals to the environment. And decomposers can also decompose decaying animals

These relations form sequences, in which each individual consumes the preceding one and is consumed by the one following, in what are called food chains or food network. In a food network, there will be fewer organisms at each level as one follows the links of the network up the chain.

These concepts lead to the idea of biomass (the total living matter in a given place), of primary productivity (the increase in the mass of plants during a given time) and of secondary productivity (the living matter produced by consumers and the decomposers in a given time).

These two last ideas are key, since they make it possible to evaluate the load capacity -- the number of organisms which can be supported by a given ecosystem. In any food network, the energy contained in the level of the producers is not completely transferred to the consumers.

And the higher one goes up the chain, the more energy and resources is lost and consumed. Thus, from an energy-and environmental-point of view, it is more efficient for humans to be primary consumers (to subsist from vegetables, grains, legumes, fruit, etc.)

than as secondary consumers (from eating herbivores, omnivores, or their products, such as milk, chickens, cattle, sheep, etc.) and still more so than as a tertiary consumer (from consuming carnivores, omnivores, or their products, such as fur, pigs, snakes, alligators, etc.). An ecosystem(s) is unstable when the load capacity is overrun and is especially unstable when a population doesn't have an ecological niche and overconsumers.

The productivity of ecosystems is sometimes estimated by comparing three types of land-based ecosystems and the total of aquatic ecosystems:

- The forests (1/3 of the Earth's land area) contain dense biomasses and are very productive. The total production of the world's forests corresponds to half of the primary production.
- Savannas, meadows, and marshes (1/3 of the Earth's land area) contain less dense biomasses, but are productive. These ecosystems represent the major part of what humans depend on for food.
- Extreme ecosystems in the areas with more extreme climates -- deserts and semi-deserts, tundra, alpine meadows, and steppes -- (1/3 of the Earth's land area) have very sparse biomasses and low productivity
- Finally, the marine and fresh water ecosystems (3/4 of Earth's surface) contain very sparse biomasses (apart from the coastal zones).

Humanity's actions over the last few centuries have seriously reduced the amount of the Earth covered by forests (deforestation), and have increased agro-ecosystems (agriculture). In recent decades, an increase in the areas occupied by extreme ecosystems has occurred (desertification).

ECOLOGICAL CRISIS AND LOSS OF ADAPTIVE CAPACITY

Generally, an ecological crisis occurs with the loss of adaptive capacity when the resilience of an environment or of a species or a population evolves in a way unfavourable to coping with perturbations that interfere with that ecosystem, landscape or species survival. It may be that the environment quality degrades compared to the species needs, after a change in an abiotic ecological factor (for example, an increase of temperature, less significant rainfalls). It may be that the environment becomes unfavourable for the survival of a species (or a population) due to an increased pressure of predation (for example overfishing). Lastly, it may be that the situation becomes unfavourable to the quality of life of the species (or the population) due to a rise in the number of individuals (overpopulation).

Ecological crises vary in length and severity, occurring within a few months or taking as long as a few million years. They can also be of natural or anthropic origin. They may relate to one unique species or to many species, as in an Extinction event. Lastly, an ecological crisis may be local (as an oil spill) or global (a rise in the sea level due to global warming).

According to its degree of endemism, a local crisis will have more or less significant consequences, from the death of many individuals to the total extinction of a species. Whatever its origin, disappearance of one or several species often will involve a rupture in the food chain, further impacting the survival of other species.

In the case of a global crisis, the consequences can be much more significant; some extinction events showed the disappearance of more than 90% of existing species at that time. However, it should be noted that the disappearance of certain species, such as the dinosaurs, by freeing an ecological niche, allowed the development and the diversification of the mammals. An ecological crisis thus paradoxically favoured biodiversity.

Sometimes, an ecological crisis can be a specific and reversible phenomenon at the ecosystem scale. But more generally, the crises impact will last.

Indeed, it rather is a connected series of events, that occur till a final point. From this stage, no return to the previous stable state is possible, and a new stable state will be set up gradually. Lastly, if an ecological crisis can cause extinction, it can also more simply reduce the quality of life of the remaining individuals.

Thus, even if the diversity of the human population is sometimes considered threatened, few people envision human disappearance at short span. However, epidemic diseases, famines, impact on health of reduction of air quality, food crises, reduction of living space, accumulation of toxic or non degradable wastes, threats on keystone species (great apes, panda, whales) are also factors influencing the well-being of people.

Due to the increases in technology and a rapidly increasing population, humans have more influence on their own environment than any other ecosystem engineer.

Some common examples of ecological crises are:

- The Exxon Valdez oil spill off the coast of Alaska in 1989
- Permian-Triassic extinction event 250 million of years ago
- Cretaceous-Tertiary extinction event 65 million years ago

- Global warming related to the Greenhouse effect. Warming could involve flooding of the Asian deltas, multiplication of extreme weather phenomena and changes in the nature and quantity of the food resources.
- Ozone layer hole issue
- Deforestation and desertification, with disappearance of many species.
- Volcanic eruptions such as Mount St. Helens and the Tunguska and other impact events
- The nuclear meltdown at Chernobyl in 1986 caused the death of many people and animals from cancer, and caused mutations in a large number of animals and people. The area around the plant is now abandoned by humans because of the large amount of radiation generated by the meltdown. Twenty years after the accident, the animals have returned.

6

Principle of Ecology and Ecosystem

The first principle of ecology is that each living organism has an ongoing and continual relationship with every other element that makes up its environment. An ecosystem can be defined as any situation where there is interaction between organisms and their environment.

The ecosystem is of two entities, the entirety of life, the biocoenosis, and the medium that life exists in, the biotope. Within the ecosystem, species are connected by food chains or food webs. Energy from the sun, captured by primary producers via photosynthesis, flows upward through the chain to primary consumers (herbivores), and then to secondary and tertiary consumers (carnivores and omnivores), before ultimately being lost to the system as waste heat. In the process, matter is incorporated into living organisms, which return their nutrients to the system via decomposition, forming biogeochemical cycles such as the carbon and nitrogen cycles.

The concept of an ecosystem can apply to units of variable size, such as a pond, a field, or a piece of dead wood. An ecosystem within another ecosystem is called a micro ecosystem. For example, an ecosystem can be a stone and all the life under it. A meso ecosystem could be a forest, and a macro ecosystem a whole eco region, with its drainage basin. The main questions when studying an ecosystem are:

- Whether the colonization of a barren area could be carried out
- Investigation the ecosystem's dynamics and changes
- The methods of which an ecosystem interacts at local, regional and global scale
- Whether the current state is stable
- Investigating the value of an ecosystem and the ways and means that interaction of ecological systems provides benefits to humans, especially in the provision of healthy water.

Ecosystems are often classified by reference to the biotopes concerned. The following ecosystems may be defined:

- As continental ecosystems, such as forest ecosystems, meadow ecosystems such as steppes or savannas, or agro-ecosystems
- As ecosystems of inland waters, such as lentic ecosystems such as lakes or ponds; or lotic ecosystems such as rivers
- As oceanic ecosystems.

Another classification can be done by reference to its communities, such as in the case of an human ecosystem.

ECOSYSTEM

A central principle of ecology is that each living organism has an ongoing and continual relationship with every other element that makes up its environment.

The sum total of interacting living organisms (the biocoenosis) and their non-living environment (the biotope) in an area is termed an *ecosystem*. Studies of ecosystems usually focus on the movement of energy and matter through the system. Almost all ecosystems run on energy captured from the sun by primary producers via photosynthesis. This energy then flows through the food chains to primary consumers (herbivores who eat and digest the plants), and on to secondary and tertiary consumers (either carnivores or omnivores). Energy is lost to living organisms when it is used by the organisms to do work, or is lost as waste heat.

Matter is incorporated into living organisms by the primary producers. Photosynthetic plants fix carbon from carbon dioxide and nitrogen from atmospheric nitrogen or nitrates present in the soil to produce amino acids. Much of the carbon and nitrogen contained in ecosystems is created by such plants, and is then consumed by secondary and tertiary consumers and incorporated into themselves.

Nutrients are usually returned to the ecosystem via decomposition. The entire movement of chemicals in an ecosystem is termed a biogeochemical cycle, and includes the carbon and nitrogen cycle. Ecosystems of any size can be studied; for example, a rock and the plant life growing on it might be considered an ecosystem. This rock might be within a plain, with many such rocks, small grass, and grazing animals— also an ecosystem. This plain might be in the tundra, which is also an ecosystem (although once they are of this size, they are generally termed ecozones or biomes). In fact, the entire terrestrial surface of the earth, all the matter which composes it, the air that is directly above it, and all the living organisms living within it can be considered as one, large ecosystem.

Ecosystems can be roughly divided into terrestrial ecosystems (including forest ecosystems, steppes, savannas, and so on), freshwater ecosystems (lakes, ponds and rivers), and marine ecosystems, depending on the dominant biotope.

Dynamics and Stability

Ecological factors that affect dynamic change in a population or species in a given ecology or environment are usually divided into two groups: abiotic and biotic. Abiotic factors are geological, geographical, hydrological, and climatological parameters. A biotope is an environmentally uniform region characterized by a particular set of abiotic ecological factors. Specific abiotic factors include:

- Water, which is at the same time an essential element to life and a milieu
- Air, which provides oxygen, nitrogen, and carbon dioxide to living species and allows the dissemination of pollen and spores
- Soil, at the same time a source of nutriment and physical support
    - o Soil pH, salinity, nitrogen and phosphorus content, ability to retain water, and density are all influential
- Temperature, which should not exceed certain extremes, even if tolerance to heat is significant for some species
- Light, which provides energy to the ecosystem through photosynthesis
- Natural disasters can also be considered abiotic.

Biocenose, or community, is a group of populations of plants, animals, microorganisms. Each population is the result of procreations between individuals of the same species and cohabitation in a given place and for a given time. When a population consists of an insufficient number of individuals, that population is threatened with extinction; the extinction of a species can approach when all biocenoses composed of individuals of the species are in decline. In small populations, consanguinity (inbreeding) can result in reduced genetic diversity, which can further weaken the biocenose. Biotic ecological factors also influence biocenose viability; these factors are considered as either intraspecific or interspecific relations. Intraspecific relations are those that are established between individuals of the same species, forming a population. They are relations of cooperation or competition, with division of the territory, and sometimes organization in hierarchical societies.

Interspecific relations—interactions between different species—are numerous, and usually described according to their beneficial, detrimental, or neutral effect (for example, mutualism (relation ++) or competition (relation —). The most significant relation is the relation of predation (to eat or to be eaten), which leads to the essential concepts in ecology of food chains (for example, the grass is consumed by the herbivore, itself consumed by a carnivore, itself consumed by a carnivore of larger size). A high predator to prey ratio can have a negative influence on both the predator and prey biocenoses in that low availability of food and high death rate prior to sexual maturity can decrease (or prevent the increase of) populations of each, respectively. Selective hunting of species by humans that leads to population decline is one example of a high predator to prey ratio in action. Other interspecific relations include parasitism, infectious disease, and competition for limited resources, which can occur when two species share the same ecological niche.

The existing interactions between the various living beings go along with a permanent mixing of mineral and organic substances, absorbed by organisms for their growth, their maintenance, and their reproduction, to be finally rejected as waste.

These permanent recycling of the elements (in particular carbon, oxygen, and nitrogen) as well as the water are called biogeochemical cycles.

They guarantee a durable stability of the biosphere (at least when unchecked human influence and extreme weather or geological phenomena are left aside). This self-regulation, supported by negative feedback controls, ensures the perenniality of the ecosystems. It is shown by the very stable concentrations of most elements of each compartment. This is referred to as homeostasis. The ecosystem also tends to evolve to a state of ideal balance, called the climax, which is reached after a succession of events (for example a pond can become a peat bog).

THE ECONOMY OF NATURAL ECOLOGICAL PROCESSES

The terms ecology and economy are rooted in the same Greek word 'oikos' or household. Yet in the context of market-oriented development they have been rendered contradictory: 'Ecological destruction is an obvious cost for economic development'-a statement which is often repeated to ecology movements. Natural resources are produced and reproduced through a complex network of ecological processes. Production is an integral part of this economy of natural ecological processes but the concepts of production and productivity in the context of development economics have been exclusively identified with the industrial production system for the market economy. Organic productivity in forestry or agriculture has also been viewed narrowly through the production of marketable products of the total productive process.

This has resulted in vast areas of resource productivity, like the production of humus by forests, or regeneration of water resources, natural evolution of genetic products, erosional production of soil fertility from parent rocks, remaining beyond the scope of economics. Many of these productive processes are dependent on a number of ecological processes. These processes are not known fully even within the natural science disciplines and economists have to make tremendous efforts to internalize them. Paradoxically, through the resource ignorant intervention of economic development at its present scale, the whole natural resource system of our planet is under threat of a serious loss of productivity in the economy of natural processes.

At present ecology movements are the sole voice to stress the economic value of these natural processes. The market-oriented development process can destroy the economy of natural processes by over exploitation of resources or by the destruction of ecological processes that are not comprehended by economic development. And these impacts are not necessarily manifested within the period of the development projects. The positive contribution of economic growth from such development may prove totally inadequate to balance the invisible or delayed negative externalities stemming from damage to the economy of natural ecological processes. In the larger context, economic growth can thus, itself become the source of underdevelopment. The ecological destruction associated with uncontrolled exploitation of natural resources for commercial gains is a symptom of the conflict between the ways of generating material wealth in the economies of-market and the natural processes. In the words of Commoner: 'Human beings have broken out of the circle of life driven not by biological needs, but the social organisation which they have devised to 'conquer' nature: means of gaining wealth which conflict with those which govern nature."

MARINE COASTAL ECOSYSTEMS

Seashores throughout the world are subject to increasing pressures from residential, recreational, and commercial development. These stresses may become more severe, for human population in the vicinity of sea-coasts is growing at twice the inland rate. Some of the pressures that we exert on coastal ecosystems are summarized in the accompanying box. All can increase the growth of algae. Among the possible consequences of disruption in almost any marine ecosystem is an increase in the opportunistic pathogens that can abet the spread of human disease, sometimes to widespread proportions. One example is cholera.

Cholera

We often think of our modern world as cleansed of the epidemic scourges of ages past. But cholera —an acute and sometimes fatal disease that is accompanied by severe diarrhea— affects more nations today than ever before. The Seventh Pandemic began when the El Tor strain left its traditional home in the Bay of Bengal in the 1960s, travelled to the east and west across Asia, and in the 1970s penetrated the continent of Africa. In 1991, the cholera pandemic reached the Americas, and during the first eighteen months more than half a million cases were reported in Latin America, with 5,000 deaths. Rapid institution of oral rehydration treatment—with clean water, sugar, and salts—limited fatalities in the Americas to about one in a hundred cases. The epidemics also had serious economic consequences. In 1991, Peru lost $770 million in seafood exports and another $250 million in lost tourist revenues because of the disease.

The microbe that transmits cholera, Vibrio cholerae, is found in a dormant or "hibernating" state in algae and microscopic animal plankton, where it can be identified using modern microbiological techniques. But once introduced to people—by drinking contaminated water or eating contaminated fish or shellfish— cholera can recycle through a population, when sewage is allowed to mix with the clean water supply.

Five years ago, in late 1992, a new strain of Vibrio cholerae—O139 Bengal—emerged in India along the coast of the Bay of Bengal. With populations unprotected by prior immunities, this hardy strain quickly spread through adjoining nations, threatening to become the agent of the world's Eighth Cholera Pandemic. For a time, in 1994, El Tor regained dominance. But by 1996, O139 Bengal had reasserted itself. The emergence of this new disease, like all others, involved the interplay of microbial, human host, and environmental factors. The largest and most intense outbreak of cholera ever recorded occurred in Rwanda in 1994, killing over 40,000 people in the space of weeks, in a nation already ravaged by civil war and ethnic strife. The tragedy of cholera in Rwanda is a reminder of the impacts of conflict and political instability on public health and biological security—just as epidemics may, in turn, contribute to political and economic stability.

Is The Ocean Warming?

Surface temperatures of the ocean have warmed this century, and a gradual warming of the deep ocean has been found in recent years in oceanographic surveys carried out in the tropical Pacific, Atlantic, and Indian Oceans, and at both poles of the Earth. These findings could be indicative of a long-term trend. Corresponding temperature measurements of the sub-surface earth, in cores drilled deep into the Arctic tundra, show a similar effect. The water that evaporates from warmer seas, and from vegetation and soils of a warmer land surface, intensifies the rate at which water cycles from ocean to clouds and back again. In so doing it increases humidity and reinforces the greenhouse effect. Warm seas are the engines that drive tropical storms and fuel the intensity of hurricanes. More high clouds can also contribute to warmer nights by trapping out-going radiation.

Some Biological Impacts

A warmer ocean can also harm marine plankton, and thus affect more advanced forms of life in the sea. A northward shift in marine flora and fauna along the California coast that has been underway since the 1930s has been associated with the long-term warming of the ocean over that span of time.

Warming—when sufficient nutrients are present—may also be contributing to the proliferation of coastal algal blooms. Harmful algal blooms of increasing extent, duration, and intensity—and involving novel, toxic species—have been reported around the world since the 1970s. Indeed, some scientists feel that the worldwide increase in coastal algal blooms may be one of the first biological signs of global environmental change.

Warm years may result in a confluence of adverse events. The 1987 El Nino was associated with the spread and new growth of tropical and temperate species of algae in higher northern and southern latitudes. Many were toxic algal blooms. In 1987, following a shoreward intrusion of Gulf Stream eddies, the dinoflagellate Gymnodimuim breve, previously found only as far north as the Gulf of Mexico, bloomed about 700 miles north, off Cape Hatteras, North Carolina, where it has since persisted, albeit at low levels. Forty-eight cases of neurological shellfish poisoning occurred in 1987, resulting in an estimated $25 million loss to the seafood industry and the local community. In the same year, anomalous rain patterns and warm Gulf Stream eddies swept unusually close to Prince Edward Island in the Gulf of St. Lawrence. The result, combined with the run-off of local pollutants after heavy rains, was a bloom of toxic diatoms. For the first time, domoic acid was produced from these algae, and then ingested by marine life. Consumption of contaminated mussels resulted in 107 instances of amnesic shellfish poisoning, from domoic acid, including three deaths and permanent, short-term memory loss in several victims.

Also in 1987, there were major losses of sea urchin and coral communities in the Caribbean, a massive sea grass die-off near the Florida Keys, and on the beaches of the North Atlantic coast, the death of numerous dolphins and other sea mammals. It has been proposed that the combination of algal toxins, chlorinated hydrocarbons like PCBs, and warming may have lowered the immunity of organisms and altered the food supply for various forms of sea life, allowing *morbilli* (measles-like) viruses to take hold.

The 1990s

For five years and eight months, from 1990 to 1995, the Pacific Ocean persisted in the warm El Nino phase, which was most unusual, for since 1877 none of these distinctive warmings had lasted more than three years. Both anomalous phases—with either warmer (El Nino) or colder (La Nina) surface waters— bring climate extremes to many regions across the globe. With the ensuing cold (La Nina) phase of 1995-1996, many regions of the world that had lived with drought during the El Nino years were now besieged with intense rains and flooding. Just as in Colombia, flooding in southern Africa was accompanied by an upsurge of vector-borne diseases, including malaria. Other areas experienced a climatic switch of the opposite kind, with drought and wildfires replacing floods. During 1996 world grain stores fell to their lowest level since the 1930s. Weather always varies; but increased variability and rapid temperature fluctuations may be a chief characteristic of our changing climate system. And increased variability and weather volatility can have significant consequences for health and for society.

Decadal Variability

The cumulative meteorological and ecological impacts of the prolonged El Nino of the early 1990s have yet to be fully evaluated, and another is now upon us. In 1995, warming in the Caribbean produced coral bleaching for the first time in Belize, as sea surface temperatures surpassed the 29°C (84°F) threshold that may damage the animal and plant tissues that make up a coral reef. In 1997, Caribbean sea surface temperatures reached 34°C (93°F) off southern Belize, and coral bleaching was accompanied by large mortalities in starfish and other sea life. Coral diseases are now sweeping through the Caribbean, and diseases that perturb marine habitat, such as coral or sea grasses, can also affect the fish stocks for which these areas serve as nurseries. A pattern of greater weather variability has begun and is expected to persist with the El Nino of 1997 and 1998. Since 1976, such anomalies in Pacific Ocean temperatures and in weather extreme events have become more frequent, more intense, and longer lasting than in the preceding 100 years, as indicated in records kept since 1877.

THE CONCEPT OF ECOSYSTEM

An ecosystem consists of the biological community that occurs in some locale, and the physical and chemical factors that make up its non-living or abiotic environment. There are many examples of ecosystems — a pond, a forest, an estuary, grassland. The boundaries are not fixed in any objective way, although sometimes they seem obvious, as with the shoreline of a small pond. Usually the boundaries of an ecosystem are chosen for practical reasons having to do with the goals of the particular study.

The study of ecosystems mainly consists of the study of certain processes that link the living, or biotic, components to the non-living, or abiotic, components. Energy transformations and biogeochemical cycling are the main processes that comprise the field of ecosystem ecology. As we learned earlier, ecology generally is defined as the interactions of organisms with one another and with the environment in which they occur. We can study ecology at the level of the individual, the population, the community, and the ecosystem.

Studies of individuals are concerned mostly about physiology, reproduction, development or behavior, and studies of populations usually focus on the habitat and resource needs of individual species, their group behaviors, population growth, and what limits their abundance or causes extinction. Studies of communities examine how populations of many species interact with one another, such as predators and their prey, or competitors that share common needs or resources.

In ecosystem ecology we put all of this together and, insofar as we can, we try to understand how the system operates as a whole. This means that, rather than worrying mainly about particular species, we try to focus on major functional aspects of the system. These functional aspects include such things as the amount of energy that is produced by photosynthesis, how energy or materials flow along the many steps in a food chain, or what controls the rate of decomposition of materials or the rate at which nutrients are recycled in the system.

The term ecosystem was coined in 1935 by the Oxford ecologist Arthur Tansley to encompass the interactions among biotic and abiotic components of the environment at a given site. The living and non-living components of an ecosystem are known as biotic and abiotic components, respectively.

Ecosystem was defined in its presently accepted form by Eugene Odom as, "an unit that includes all the organisms, i.e., the community in a given area interacting with the physical environment so that a flow of energy leads to clearly defined tropic structure, biotic diversity and material cycles, i.e., exchange of materials between living and non-living, within the system".

Smith (1966) has summarized common characteristics of most of the ecosystems as follows:

1. The ecosystem is a major structural and functional unit of ecology.
2. The structure of an ecosystem is related to its species diversity in the sense that complex ecosystem have high species diversity.
3. The function of ecosystem is related to energy flow and material cycles within and outside the system.
4. The relative amount of energy needed to maintain an ecosystem depends on its structure. Complex ecosystems needed less energy to maintain them.
5. Young ecosystems develop and change from fewer complexes to more complex ecosystems, through the process called succession.

6. Each ecosystem has its own energy budget, which cannot be exceeded.
7. Adaptation to local environmental conditions is the important feature of the biotic components of an ecosystem, failing which they might perish.
8. The function of every ecosystem involves a series of cycles, e.g., water cycle, nitrogen cycle, oxygen cycle, etc. these cycles are driven by energy. A continuation or existence of ecosystem demands exchange of materials/nutrients to and from the different components.

ECOSYSTEM PRODUCTIVITY

In an ecosystem, the connections between species are generally related to their role in the food chain. There are three categories of organisms:

- *Producers* or *Autotrophs* — Usually plants or cyanobacteria that are capable of photosynthesis but could be other organisms such as the bacteria near ocean vents that are capable of chemosynthesis.
- *Consumers* or *Heterotrophs* — Animals, which can be primary consumers (herbivorous), or secondary or tertiary consumers (carnivorous and omnivores).
- *Decomposers* or *Detritivores* — Bacteria, fungi, and insects which degrade organic matter of all types and restore nutrients to the environment. The producers will then consume the nutrients, completing the cycle.

These relations form sequences, in which each individual consumes the preceding one and is consumed by the one following, in what are called food chains or food networks. In a food network, there will be fewer organisms at each level as one follows the links of the network up the chain, forming a pyramid.

These concepts lead to the idea of biomass (the total living matter in an ecosystem), primary productivity (the increase in organic compounds), and secondary productivity (the living matter produced by consumers and the decomposers in a given time).

These last two ideas are key, since they make it possible to evaluate the carrying capacity — the number of organisms that can be supported by a given ecosystem. In any food network, the energy contained in the level of the producers is not completely transferred to the consumers.

The higher up the chain, the more energy and resources are lost.

Thus, from a purely energy and nutrient point of view, it is more efficient for humans to be primary consumers (to subsist from vegetables, grains, legumes, fruit, etc.) than to be secondary consumers (consuming herbivores, omnivores, or their products) and still more so than as a tertiary consumer (consuming carnivores, omnivores, or their products). An ecosystem is unstable when the carrying capacity is overrun. The total productivity of ecosystems is sometimes estimated by comparing three types of land-based ecosystems and the total of aquatic ecosystems. Slightly over half of primary production is estimated to occur on land, and the rest in the ocean.

- The forests (1/3 of the Earth's land area) contain dense biomasses and are very productive.
- Savannas, meadows, and marshes (1/3 of the Earth's land area) contain less dense biomasses, but are productive. These ecosystems represent the major part of what humans depend on for food.
- Extreme ecosystems in the areas with more extreme climates — deserts and semi-deserts, tundra, alpine meadows, and steppes — (1/3 of the Earth's land area) have very sparse biomasses and low productivity
- Finally, the marine and fresh water ecosystems (3/4 of Earth's surface) contain very sparse biomasses (apart from the coastal zones).

Ecosystems differ in biomass (grams carbon per square meter) and productivity (grams carbon per square meter per day), and direct comparisons of biomass and productivity may not be valid. An ecosystem such as that found in taiga may be high in biomass, but slow growing and thus low in productivity. Ecosystems are often compared on the basis of their turnover (production ratio) or turnover time which is the reciprocal of turnover.

Humanity's actions over the last few centuries have seriously reduced the amount of the Earth covered by forests (deforestation), and have increased agro-ecosystems. In recent decades, an increase in the areas occupied by extreme ecosystems has occurred, such as desertification.

Ecological Crisis

Generally, an ecological crisis occurs with the loss of adaptive capacity when the resilience of an environment or of a species or a population evolves in a way unfavourable to coping with perturbations that interfere with that ecosystem, landscape or species survival. It may be that the environment becomes unfavourable for the survival of a species (or a population) due to an increased pressure of predation (for example overfishing). Lastly, it may be that the situation becomes unfavourable to the quality of life of the species (or the population) due to a rise in the number of individuals (overpopulation). Ecological crises vary in length and severity, occurring within a few months or taking as long as a few million years. They can also be of natural or anthropic origin. They may relate to one unique species or to many species, as in an Extinction event. Lastly, an ecological crisis may be local (as an oil spill) or global (a rise in the sea level due to global warming).

According to its degree of endemism, a local crisis will have more or less significant consequences, from the death of many individuals to the total extinction of a species. Whatever its origin, disappearance of one or several species often will involve a rupture in the food chain, further impacting the survival of other species. In the case of a global crisis, the consequences can be much more significant; some extinction events showed the disappearance of more than 90% of existing species at that time. However, it should be noted that the disappearance of certain species, such as the dinosaurs, by freeing an ecological niche, allowed the development and the diversification of the mammals. An ecological crisis thus paradoxically favoured biodiversity.

Sometimes, an ecological crisis can be a specific and reversible phenomenon at the ecosystem scale. But more generally, the crises impact will last. Indeed, it rather is a connected series of events, that occur till a final point. From this stage, no return to the previous stable state is possible, and a new stable state will be set up gradually.

Lastly, if an ecological crisis can cause extinction, it can also more simply reduce the quality of life of the remaining individuals. Thus, even if the diversity of the human population is sometimes considered threatened, few people envision human disappearance at short span. However, epidemic diseases, famines, impact on health of reduction of air quality, food crises, reduction of living space, accumulation of toxic or non degradable wastes, threats on keystone species (great apes, panda, whales) are also factors influencing the well-being of people. Due to the increases in technology and a rapidly increasing population, humans have more influence on their own environment than any other ecosystem engineer.

DYNAMIC OF ECOLOGICAL SYSTEMS

In terrestrial ecosystems, the earlier timing of spring events, and poleward and upward shifts in plant and animal ranges, have been linked with high confidence to recent warming. Future climate change is expected to particularly affect certain ecosystems, including tundra, mangroves, and coral reefs. It is expected that most ecosystems will be affected by higher atmospheric CO_2 levels, combined with higher global temperatures. Overall, it is expected that climate change will result in the extinction of many species and reduced diversity of ecosystems.

Biodiversity

Deforestation on a human scale results in declines in biodiversity and on a natural global scale is known to cause the extinction of many species. The removal or destruction of areas of forest cover has resulted in a degraded environment with reduced biodiversity. Forests support biodiversity, providing habitat for wildlife; moreover, forests foster medicinal conservation. With forest biotopes being irreplaceable source of new drugs (such as taxol), deforestation can destroy genetic variations (such as crop resistance) irretrievably. Since the tropical rainforests are the most diverse ecosystems on Earth and about 80 per cent of the world's known biodiversity could be found in tropical rainforests, removal or destruction of significant areas of forest cover has resulted in a degraded environment with reduced biodiversity.

It has been estimated that we are losing 137 plant, animal and insect species every single day due to rainforest deforestation, which equates to 50,000 species a year. Others state that tropical rainforest deforestation is contributing to the ongoing Holocene mass extinction. The known extinction rates from deforestation rates are very low, approximately 1 species per year from mammals and birds which extrapolates to approximately 23,000 species per year for all species. Predictions have been made that more than 40 per cent of the animal and plant species in Southeast Asia could be wiped out in the 21st century. Such predictions were called into question by 1995 data that show that within regions of Southeast Asia much of the original forest has been converted to monospecific plantations, but that potentially endangered species are few and tree flora remains widespread and stable.

Scientific understanding of the process of extinction is insufficient to accurately make predictions about the impact of deforestation on biodiversity. Most predictions of forestry related biodiversity loss are based on species-area models, with an underlying assumption that as the forest declines species diversity will decline similarly. However, many such models have been proven to be wrong and loss of habitat does not necessarily lead to large scale loss of species. Species-area models are known to overpredict the number of species known to be threatened in areas where actual deforestation is ongoing, and greatly overpredict the number of threatened species that are widespread.

Hydrological

The water cycle is also affected by deforestation. Trees extract groundwater through their roots and release it into the atmosphere. When part of a forest is removed, the trees no longer evaporate away this water, resulting in a much drier climate. Deforestation reduces the content of water in the soil and groundwater as well as atmospheric moisture. The dry soil leads to lower water intake for the trees to extract. Deforestation reduces soil cohesion, so that erosion, flooding and landslides ensue. Shrinking forest cover lessens the landscape's capacity to intercept, retain and transpire precipitation. Instead of trapping precipitation, which then percolates to groundwater systems, deforested areas become sources of surface water runoff, which moves much faster than subsurface flows. That quicker transport of surface water can translate into flash flooding and more localised floods than would occur with the forest cover. Deforestation also contributes to decreased evapotranspiration, which lessens atmospheric moisture which in some cases affects precipitation levels downwind from the deforested area, as water is not recycled to downwind forests, but is lost in runoff and returns directly to the oceans. According to one study, in deforested north and northwest China, the average annual precipitation decreased by one third between the 1950s and the 1980s.

Trees, and plants in general, affect the water cycle significantly:

- their canopies intercept a proportion of precipitation, which is then evaporated back to the atmosphere (canopy interception);
- their litter, stems and trunks slow down surface runoff;
- their roots create macropores – large conduits – in the soil that increase infiltration of water;
- they contribute to terrestrial evaporation and reduce soil moisture via transpiration;
- their litter and other organic residue change soil properties that affect the capacity of soil to store water.
- their leaves control the humidity of the atmosphere by transpiring. 99 per cent of the water absorbed by the roots moves up to the leaves and is transpired.

As a result, the presence or absence of trees can change the quantity of water on the surface, in the soil or groundwater, or in the atmosphere. This in turn changes erosion rates and the availability of water for either ecosystem functions or human services. The forest may have little impact on flooding in the case of large rainfall events, which overwhelm the storage capacity of forest soil if the soils are at or close to saturation. Tropical rainforests produce about 30 per cent of our planet's fresh water.

Soil

Undisturbed forests have a very low rate of soil loss, approximately 2 metric tons per square kilometre (6 short tons per square mile). Deforestation generally increases rates of soil erosion, by increasing the amount of runoff and reducing the protection of the soil from tree litter. This can be an advantage in excessively leached tropical rain forest soils. Forestry operations themselves also increase erosion through the development of roads and the use of mechanised equipment. China's Loess Plateau was cleared of forest millennia ago. Since then it has been eroding, creating dramatic incised valleys, and providing the sediment that gives the Yellow River its yellow colour and that causes the flooding of the river in the lower reaches (hence the river's nickname 'China's sorrow').

Removal of trees does not always increase erosion rates. In certain regions of southwest US, shrubs and trees have been encroaching on grassland. The trees themselves enhance the loss of grass between tree canopies. The bare intercanopy areas become highly erodible. The US Forest Service, in Bandelier National Monument for example, is studying how to restore the former ecosystem, and reduce erosion, by removing the trees. Tree roots bind soil together, and if the soil is sufficiently shallow they act to keep the soil in place by also binding with underlying bedrock. Tree removal on steep slopes with shallow soil thus increases the risk of landslides, which can threaten people living nearby. However most deforestation only affects the trunks of trees, allowing for the roots to stay rooted, negating the landslide.

Forest Transition Theory

The forest area change may follow a pattern suggested by the forest transition (FT) theory, whereby at early stages in its development a country is characterised by high forest cover and low deforestation rates (HFLD countries).

Then deforestation rates accelerate (HFHD, high forest cover – high deforestation rate), and forest cover is reduced (LFHD. low forest cover – high deforestation rate), before the deforestation rate slows (LFLD, low forest cover – low deforestation rate), after which forest cover stabilises and eventually starts recovering. FT is not a "law of nature," and the pattern is influenced by national context (for example, human population density, stage of development, structure of the economy), global economic forces, and government policies. A country may reach very low levels of forest cover before it stabilises, or it might through good policies be able to "bridge" the forest transition.

FT depicts a broad trend, and an extrapolation of historical rates therefore tends to underestimate future BAU deforestation for counties at the early stages in the transition (HFLD), while it tends to overestimate BAU deforestation for countries at the later stages (LFHD and LFLD).

Countries with high forest cover can be expected to be at early stages of the FT. GDP per capita captures the stage in a country's economic development, which is linked to the pattern of natural resource use, including forests. The choice of forest cover and GDP per capita also fits well with the two key scenarios in the FT:

(i) a forest scarcity path, where forest scarcity triggers forces (for example, higher prices of forest products) that lead to forest cover stabilisation; and

(ii) an economic development path, where new and better off-farm employment opportunities associated with economic growth (= increasing GDP per capita) reduce profitability of frontier agriculture and slows deforestation.

Historical Causes

Prehistory: The Carboniferous Rainforest Collapse, was an event that occurred 300 million years ago. Climate change devastated tropical rainforests causing the extinction of many plant and animal species. The change was abrupt, specifically, at this time climate became cooler and drier, conditions that are not favourable to the growth of rainforests and much of the biodiversity within them. Rainforests were fragmented forming shrinking 'islands' further and further apart.

This sudden collapse affected several large groups, effects on amphibians were particularly devastating, while reptiles fared better, being ecologically adapted to the drier conditions that followed. Rainforests once covered 14 per cent of the earth's land surface; now they cover a mere 6 per cent and experts estimate that the last remaining rainforests could be consumed in less than 40 years. Small scale deforestation was practiced by some societies for tens of thousands of years before the beginnings of civilization.

The first evidence of deforestation appears in the Mesolithic period. It was probably used to convert closed forests into more open ecosystems favourable to game animals. With the advent of agriculture, larger areas began to be deforested, and fire became the prime tool to clear land for crops. In Europe there is little solid evidence before 7000 BC.

Mesolithic foragers used fire to create openings for red deer and wild boar. In Great Britain, shade-tolerant species such as oak and ash are replaced in the pollen record by hazels, brambles, grasses and nettles. Removal of the forests led to decreased transpiration, resulting in the formation of upland peat bogs. Widespread decrease in elm pollen across Europe between 8400–8300 BC and 7200–7000 BC, starting in southern Europe and gradually moving north to Great Britain, may represent land clearing by fire at the onset of Neolithic agriculture.

The Neolithic period saw extensive deforestation for farming land. Stone axes were being made from about 3000 BC not just from flint, but from a wide variety of hard rocks from across Britain and North America as well. They include the noted Langdale axe industry in the English Lake District, quarries developed at Penmaenmawr in North Wales and numerous other locations. Rough-outs were made locally near the quarries, and some were polished locally to give a fine finish. This step not only increased the mechanical strength of the axe, but also made penetration of wood easier. Flint was still used from sources such as Grimes Graves but from many other mines across Europe. Evidence of deforestation has been found in Minoan Crete; for example the environs of the Palace of Knossos were severely deforested in the Bronze Age.

Pre-industrial History

Throughout most of history, humans were hunter gatherers who hunted within forests. In most areas, such as the Amazon, the tropics, Central America, and the Caribbean, only after shortages of wood and other forest products occur are policies implemented to ensure forest resources are used in a sustainable manner.

In ancient Greece, Tjeered van Andel and co-writers summarised three regional studies of historic erosion and alluviation and found that, wherever adequate evidence exists, a major phase of erosion follows, by about 500-1,000 years the introduction of farming in the various regions of Greece, ranging from the later Neolithic to the Early Bronze Age. The thousand years following the mid-first millennium BCE saw serious, intermittent pulses of soil erosion in numerous places. The historic silting of ports along the southern coasts of Asia Minor (*e.g.* Clarus, and the examples of Ephesus, Priene and Miletus, where harbors had to be abandoned because of the silt deposited by the Meander) and in coastal Syria during the last centuries BC.

Easter Island has suffered from heavy soil erosion in recent centuries, aggravated by agriculture and deforestation. Jared Diamond gives an extensive look into the collapse of the ancient Easter Islanders in his book *Collapse*. The disappearance of the island's trees seems to coincide with a decline of its civilization around the 17th and 18th century. He attributed the collapse to deforestation and over-exploitation of all resources.

The famous silting up of the harbor for Bruges, which moved port commerce to Antwerp, also followed a period of increased settlement growth (and apparently of deforestation) in the upper river basins. In early medieval Riez in upper Provence, alluvial silt from two small rivers raised the riverbeds and widened the floodplain, which slowly buried the Roman settlement in alluvium and gradually moved new construction to higher ground; concurrently the headwater valleys above Riez were being opened to pasturage. A typical progress trap was that cities were often built in a forested area, which would provide wood for some industry (for example, construction, shipbuilding, pottery). When deforestation occurs without proper replanting, however; local wood supplies become difficult to obtain near enough to remain competitive, leading to the city's abandonment, as happened repeatedly in Ancient Asia Minor. Because of fuel needs, mining and metallurgy often led to deforestation and city abandonment.

With most of the population remaining active in (or indirectly dependent on) the agricultural sector, the main pressure in most areas remained land clearing for crop and cattle farming. Enough wild green was usually left standing (and partially used, for example, to collect firewood, timber and fruits, or to graze pigs) for wildlife to remain viable. The elite's (nobility and higher clergy) protection of their own hunting privileges and game often protected significant woodlands.

Major parts in the spread (and thus more durable growth) of the population were played by monastical 'pioneering' (especially by the Benedictine and Commercial orders) and some feudal lords' recruiting farmers to settle (and become tax payers) by offering relatively good legal and fiscal conditions.

Even when speculators sought to encourage towns, settlers needed an agricultural belt around or sometimes within defensive walls.

When populations were quickly decreased by causes such as the Black Death or devastating warfare (for example, Genghis Khan's Mongol hordes in eastern and central Europe, Thirty Years' War in Germany), this could lead to settlements being abandoned. The land was reclaimed by nature, but the secondary forests usually lacked the original biodiversity.

From 1100 to 1500 AD, significant deforestation took place in Western Europe as a result of the expanding human population.

The large-scale building of wooden sailing ships by European (coastal) naval owners since the 15th century for exploration, colonisation, slave trade– and other trade on the high seas consumed many forest resources.

Piracy also contributed to the over harvesting of forests, as in Spain. This led to a weakening of the domestic economy after Columbus' discovery of America, as the economy became dependent on colonial activities (plundering, mining, cattle, plantations, trade, etc.)

In *Changes in the Land* (1983), William Cronon analysed and documented 17th-century English colonists' reports of increased seasonal flooding in New England during the period when new settlers initially cleared the forests for agriculture. They believed flooding was linked to widespread forest clearing upstream.

The massive use of charcoal on an industrial scale in Early Modern Europe was a new type of consumption of western forests; even in Stuart England, the relatively primitive production of charcoal has already reached an impressive level.

Stuart England was so widely deforested that it depended on the Baltic trade for ship timbers, and looked to the untapped forests of New England to supply the need.

Each of Nelson's Royal Navy war ships at Trafalgar (1805) required 6,000 mature oaks for its construction. In France, Colbert planted oak forests to supply the French navy in the future. When the oak plantations matured in the mid-19th century, the masts were no longer required because shipping had changed.

Norman F. Cantor's summary of the effects of late medieval deforestation applies equally well to Early Modern Europe:

> *Europeans had lived in the midst of vast forests throughout the earlier medieval centuries. After 1250 they became so skilled at deforestation that by 1500 they were running short of wood for heating and cooking. They were faced with a nutritional decline because of the elimination of the generous supply of wild game that had inhabited the now-disappearing forests, which throughout medieval times had provided the staple of their carnivorous high-protein diet. By 1500 Europe was on the edge of a fuel and nutritional disaster [from] which it was saved in the sixteenth century only by the burning of soft coal and the cultivation of potatoes and maize.*

7

Air Pollution: Causes and Effects

AIR POLLUTION

Humans probably first experienced harm from air pollution when they built fires in poorly ventilated caves. Since then we have gone on to pollute more of the earth's surface. Until recently, environmental pollution problems have been local and minor because of the Earth's own ability to absorb and purify minor quantities of pollutants. The industrialization of society, the introduction of motorized vehicles, and the explosion of the population, are factors contributing toward the growing air pollution problem. At this time it is urgent that we find methods to clean up the air.

The primary air pollutants found in most urban areas are carbon monoxide, nitrogen oxides, sulphur oxides, hydrocarbons, and particulate matter (both solid and liquid). These pollutants are dispersed throughout the world's atmosphere in concentrations high enough to gradually cause serious health problems. Serious health problems can occur quickly when air pollutants are concentrated, such as when massive injections of sulphur dioxide and suspended particulate matter are emitted by a large volcanic eruption.

Air Pollution in the Home

You cannot escape air pollution, not even in your own home. "In 1985 the Environmental Protection Agency (EPA) reported that toxic chemicals found in the air of almost every American home are three times more likely to cause some type of cancer than outdoor air pollutants". The health problems in these buildings are called "sick building syndrome". "An estimated one-fifth to one-third of all U.S. buildings are now considered "sick".

The EPA has found that the air in some office buildings is 100 times more polluted than the air outside. Poor ventilation causes about half of the indoor air pollution problems. The rest come from specific sources such as copying machines, electrical and telephone cables, mold and microbe-harboring air conditioning systems and ducts, cleaning fluids, cigarette smoke, carpet, latex caulk and paint, vinyl molding, linoleum tile, and building materials and furniture that emit air pollutants such as formaldehyde.

A major indoor air pollutant is radon-222, a colorless, odorless, tasteless, naturally occurring radioactive gas produced by the radioactive decay of uranium-238.

"According to studies by the EPA and the National Research Council, exposure to radon is second only to smoking as a cause of lung cancer". Radon enters through pores and cracks in concrete when indoor air pressure is less than the pressure of gasses in the soil. Indoor air will be healthier than outdoor air if you use an energy recovery ventilator to provide a consistent supply of fresh filtered air and then seal air leaks in the shell of your home.

Sources of Pollutants

The two main sources of pollutants in urban areas are transportation (predominantly automobiles) and fuel combustion in stationary sources, including residential, commercial, and industrial heating and cooling and coal-burning power plants. Motor vehicles produce high levels of carbon monoxides (CO) and a major source of hydrocarbons (HC) and nitrogen oxides (NOx). Whereas, fuel combustion in stationary sources is the dominant source of sulphur dioxide (SO_2).

Carbon Dioxide

Carbon dioxide (CO_2) is one of the major pollutants in the atmosphere. Major sources of CO_2 are fossil fuels burning and deforestation.

"The concentrations of CO_2 in the air around 1860 before the effects of industrialization were felt, is assumed to have been about 290 parts per million (ppm). In the hundred years and more since then, the concentration has increased by about 30 to 35 ppm that is by 10 percent". Industrial countries account for 65% of CO_2 emissions with the United States and Soviet Union responsible for 50%. Less developed countries (LDCs), with 80% of the world's people, are responsible for 35% of CO_2 emissions but may contribute 50% by 2020. "Carbon dioxide emissions are increasing by 4% a year".

In 1975, 18 thousand million tons of carbon dioxide (equivalent to 5 thousand million tons of carbon) were released into the atmosphere, but the atmosphere showed an increase of only 8 billion tons (equivalent to 2.2 billion tons of carbon". The ocean waters contain about sixty times more CO_2 than the atmosphere. If the equilibrium is disturbed by externally increasing the concentration of CO_2 in the air, then the oceans would absorb more and more CO_2. If the oceans can no longer keep pace, then more CO_2 will remain into the atmosphere. As water warms, its ability to absorb CO_2 is reduced.

CO_2 is a good transmitter of sunlight, but partially restricts infrared radiation going back from the earth into space. This produces the so-called greenhouse effect that prevents a drastic cooling of the Earth during the night.

Increasing the amount of CO_2 in the atmosphere reinforces this effect and is expected to result in a warming of the Earth's surface. Currently carbon dioxide is responsible for 57% of the global warming trend. Nitrogen oxides contribute most of the atmospheric contaminants.

N0X-nitric oxide (N0) and nitrogen dioxide ($N0_2$):

- Natural component of the Earth's atmosphere.
- Important in the formation of both acid precipitation and photochemical smog (ozone), and causes nitrogen loading.
- Comes from the burning of biomass and fossil fuels.
- 30 to 50 million tons per year from human activities, and natural 10 to 20 million tons per year.
- Average residence time in the atmosphere is days.
- Has a role in reducing stratospheric ozone.

N20-nitrous oxide:

- Natural component of the Earth's atmosphere.
- Important in the greenhouse effect and causes nitrogen loading.
- Human inputs 6 million tons per year, and 19 million tons per year by nature.
- Residence time in the atmosphere about 170 years.
- 1700 (285 parts per billion), 1990 (310 parts per billion), 2030 (340 parts per billion).
- Comes from nitrogen based fertilizers, deforestation, and biomass burning.

Sulphur and Chlorofluorocarbons (CFCs)

Sulphur dioxide is produced by combustion of sulphur-containing fuels, such as coal and fuel oils. Also, in the process of producing sulfuric acid and in metallurgical process involving ores that contain sulphur.

Sulphur oxides can injure man, plants and materials. At sufficiently high concentrations, sulphur dioxide irritates the upper respiratory tract of human beings because potential effect of sulphur dioxide is to make breathing more difficult by causing the finer air tubes of the lung to constrict. "Power plants and factories emit 90% to 95% of the sulphur dioxide and 57% of the nitrogen oxides in the United States. Almost 60% of the SO_2 emissions are released by tall smoke stakes, enabling the emissions to travel long distances".

As emissions of sulphur dioxide and nitric oxide from stationary sources are transported long distances by winds, they form secondary pollutants such as nitrogen dioxide, nitric acid vapour, and droplets containing solutions of sulfuric acid, sulfate, and nitrate salts. These chemicals descend to the earth's surface in wet form as rain or snow and in dry form as a gases fog, dew, or solid particles. This is known as acid deposition or acid rain.

Chlorofluorocarbons (CFCs)

CFCs are lowering the average concentration of ozone in the stratosphere. "Since 1978 the use of CFCs in aerosol cans has been banned in the United States, Canada, and most Scandinavian countries. Aerosols are still the largest use, accounting for 25% of global CFC use". Spray cans, discarded or leaking refrigeration and air conditioning equipment, and the burning plastic foam products release the CFCs into the atmosphere. Depending on the type, CFCs stay in the atmosphere from 22 to 111 years. Chlorofluorocarbons move up to the stratosphere gradually over several decades. Under high energy ultra violet (UV) radiation, they break down and release chlorine atoms, which speed up the breakdown of ozone (O_3) into oxygen gas (O_2).

Chlorofluorocarbons, also known as Freons, are greenhouse gases that contribute to global warming. Photochemical air pollution is commonly referred to as "smog". Smog, a contraction of the words smoke and fog, has been caused throughout recorded history by water condensing on smoke particles, usually from burning coal. With the introduction of petroleum to replace coal economies in countries, photochemical smog has become predominant in many cities, which are located in sunny, warm, and dry climates with many motor vehicles. The worst episodes of photochemical smog tend to occur in summer.

Smog

Photochemical smog is also appearing in regions of the tropics and subtropics where savanna grasses are periodically burned. Smog's unpleasant properties result from the irradiation by sunlight of hydrocarbons caused primarily by unburned gasoline emitted by automobiles and other combustion sources. The products of photochemical reactions includes organic particles, ozone, aldehydes, ketones, peroxyacetyl nitrate, organic acids, and other oxidants. Ozone is a gas created by nitrogen dioxide or nitric oxide when exposed to sunlight. Ozone causes eye irritation, impaired lung function, and damage to trees and crops. Another form of smog is called industrial smog.

This smog is created by burning coal and heavy oil that contain sulphur impurities in power plants, industrial plants, etc... The smog consists mostly of a mixture of sulphur dioxide and fog. Suspended droplets of sulfuric acid are formed from some of the sulphur dioxide, and a variety of suspended solid particles. This smog is common during the winter in cities such as London, Chicago, Pittsburgh. When these cities burned large amounts of coal and heavy oil without control of the output, large-scale problems were witnessed. In 1952 London, England, 4,000 people died as a result of this form of fog. Today coal and heavy oil are burned only in large boilers and with reasonably good control or tall smokestacks so that industrial smog is less of a problem. However, some countries such as China, Poland, Czechoslovakia, and some other eastern European countries, still burn large quantities of coal without using adequate controls.

Pollution Damage to Plants

With the destruction and burning of the rain forests more and more CO_2 is being released into the atmosphere. Trees play an important role in producing oxygen from carbon dioxide. "A 115 year old Beech tree exposes about 200,000 leaves with a total surface to 1200 square meters. During the course of one sunny day such a tree inhales 9,400 liters of carbon dioxide to produce 12 kilograms of carbohydrate, thus liberating 9,400 liters of oxygen. Through this mechanism about 45,000 liters of air are regenerated which is sufficient for the respiration of 2 to 3 people". This process is called photosynthesis which all plants go though but some yield more and some less oxygen. As long as no more wood is burnt than is reproduced by the forests, no change in atmospheric CO_2 concentration will result.

Pollutants such as sulphur dioxide, nitrogen oxides, ozone and peroxyacl nitrates (PANs), cause direct damage to leaves of crop plants and trees when they enter leaf pores (stomates). Chronic exposure of leaves and needles to air pollutants can also break down the waxy coating that helps prevent excessive water loss and damage from diseases, pests, drought and frost. "In the midwestern United States crop losses of wheat, corn, soybeans, and peanuts from damage by ozone and acid deposition amount to about $5 billion a year".

Reducing Pollution

You can help to reduce global air pollution and climate change by driving a car that gets at least 35 miles a gallon, walking, bicycling, and using mass transit when possible. Replace incandescent light bulbs with compact fluorescent bulbs, make your home more energy efficient, and buy only energy efficient appliances. Recycle newspapers, aluminum, and other materials.

Plant trees and avoid purchasing products such as Styrofoam that contain CFCs. Support much stricter clean air laws and enforcement of international treaties to reduce ozone depletion and slow global warming.

Earth is everybody's home and nobody likes living in a dirty home. Together, we can make the earth a cleaner, healthier and more pleasant place to live.

BACKGROUND AND INTRODUCTION TO AIR POLLUTION

Air quality is affected by economic activities which introduces pollutants into the atmosphere that pose threats to human health and other life forms on earth. It furthermore has the potential to change the climate with unpredictable, but potentially severe consequences on a local and global scale. Because large bodies of air cannot be contained, atmospheric pollution can only be controlled at its source.

At present there is no comprehensive information on air quality or on the levels of emissions entering the atmosphere from different sources. Major areas of concern are high levels of smoke and other pollutants in poorer urban and rural households without electricity, and the impacts of the mining, energy, mineral and petro-chemical industries on air quality standards.

Air pollution is a major environmental problem throughout the whole of South Africa. South Africa derives 75,2 % of its energy from coal (a non-renewable resource), and most air pollution problems thus result from man's pattern of energy use and production. The rest of the energy comes from the following sources: 10,1% from crude oil, 9,8% from renewable bagasse and wood, 3,1% from nuclear power, 1,6% from gas and 0,2% from hydro power.

The worst levels of air pollution in South Africa is found in the Eastern Highveld of Mpumalanga (formerly the Eastern Transvaal). It covers an area of 30 000 km^2 and is home to ten ESKOM power stations, of which five are the largest in the world. The three main power stations, Matla, Duvha and Arnot produce 860 tons of SO_2 per km^2 per year. The area also contains coal mines, Sasol petrochemical plants and other industries. The major dust dome in South Africa is the Vaal Triangle to the south of Gauteng.

The worst polluted areas in Greater Johannesburg are the surrounding Highveld areas due to the combustion of fuels for the generation of electricity, and Soweto, due to the burning of coal for heating and cooking. In the winter, smoke and SO_2 from the townships are the main forms of air pollution, while vehicles and industries contribute to air pollution throughout the year.

Factors influencing the pollution problem in South Africa and Greater Johannesburg :

Seasonal Patterns

Unstable air circulates and dissipates pollutants in summer due to the low pressure over the land. In winter there is a high pressure over the country and pollutants are trapped in stable air and not dissipated or transported elsewhere.

Temperature Inversions

In the winter warm air rises from artificially heated cites or the sides of valleys. Cold night air moves in below the hot air, and temperature thus rises with height, called a low-level inversion. Pollutants are trapped in the cold layer by the warmer air above and can not be dissipated. Low level inversion of hundreds of meters deep commonly occur over Johannesburg in the winter. Even if the pollutants manage to escape the low-level inversion they still become trapped in high level inversions, which occur when cooler rural air moves in beneath warmer city air. These inversions commonly occur over Greater Johannesburg at a height of 1 200 – 1 600m above the ground.

Height above Sea Level

Due to Greater Johannesburg being about 1 600 – 2000 m above sea level, the levels of oxygen on the Highveld are 20% less than that at the coast. This means that incomplete combustion of fossil fuels takes place.

Wind speed and direction influences the rate of diffusion of pollutants. The following table gives the average wind speeds (m/s) and direction at the Johannesburg International Airport for 1993-1998 (Weather Bureau, 1999). The prevailing wind on the reef in Greater Johannesburg is north-northwest, especially in the wintertime. The wind can turn around and blow from the southeast in summer when it brings rain.

CLIMATE CHANGE

The Earths atmosphere keeps the planet warm. Without the warming cover of natural greenhouse gases, mainly carbon dioxide (CO_2) and water vapour, life could not exist on Earth. Through the release of greenhouse gases such as CO_2, methane, CFCs and N_2O caused by human activities, our climate will change. How fast, and where exactly, is still controversial, but there is consensus in the scientific community that the consequences may be serious:

- The expected rise in sea levels may threaten islands and nations with low coast lines;
- Changes in rainfall levels and patterns may affect natural vegetation, agriculture and forestry;
- The loss of biodiversity may be accelerated if climate zones move so fast that species (e.g. in rain forests) cannot follow them;
- Weather anomalies such as hurricanes may occur more frequently, causing immense damage to humans and their property, and to nature.

Not all possible consequences are fully understood. For example, it is very uncertain:

- To what extent greenhouse gas-induced disturbances of the ocean-atmosphere equilibrium contribute to altered global circulation patterns such as the El Niño phenomenon;

- Whether the gulf stream, Europes central heating, could change its direction and/or intensity, thus leading to a drastic cooling of Europes climate;

The Member States of the European Union are responsible for about one quarter of the global anthropogenic greenhouse gas emissions. A global strategy must also take into account the emissions caused by rapid economic growth in developing countries.

Most of these nations are technology buyers using, voluntarily or forced by the markets, energy-relevant technologies such as cars and power plants that are developed and produced in Europe, Japan and the USA. Progress towards more energy-efficient technologies in the EU may therefore have strong positive side-effects in other regions of the globe that may even be greater, in the long run, than the direct emission reductions observed in the EU itself.

At the 1992 United Nations Conference on Environment and Development held in Rio de Janeiro, the Framework Convention on Climate Change was adopted as the basis for global political action. Under this convention, new commitments to reduce emissions of greenhouse gases beyond the year 2000 were agreed in Kyoto in December 1997. The Kyoto Protocol, adopted by consensus by some 150 parties, stipulates that Annex 1 Parties (mainly industrialised countries) shall individually or jointly reduce their aggregate emissions of a "basket" of six greenhouse gases to 5% below 1990 levels in the period 2008-2012.

In contrast to this political target, the scientific community speaking through the voice of the Intergovernmental Panel on Climate Change (IPCC) demands a much more substantial three-quarter reduction of the current greenhouse gas emissions. This would require an almost complete phasing out of fossil fuels, with significant (but overall not necessarily negative) consequences for Europes economies.

For its part the EU undertook to reduce by 8% over the next 15 years emissions that cause global warming, compared with 7% for USA and 6% for Japan and Canada. The reduction target is calculated using the Greenhouse Warming Potentials (GWP) developed by the IPCC as "weighting coefficients", and thus is perhaps a first example how a future "Pressure Index Climate Change" could be used to monitor the implementation of the Kyoto process on the basis of one simple statistical figure.

The following indicators show how far the European Union has been successful in reducing greenhouse gas emissions. NOx and SO_2, although not greenhouse gases, have been added following the recommendations of the Scientific Advisory Groups (SAG) on Climate Change, because they play an important role in the scientific interpretation of global temperature trends.

Loss of Biodiversity

The Global Convention on Biological Diversity, signed in 1992 at the Earth Summit, describes biodiversity as the "variability among all living organisms from all sources, including terrestrial, marine and other aquatic ecosystems and ecological complexes of which they are part, this includes diversity within species, between species and of ecosystems."

Among the ten policy fields, Loss of Biodiversity is probably the most controversial one. The diversity of nature is the result of an evolutionary process that started about two billion years ago. When looking, for example, at the destruction of rain forests over the last twenty years, it becomes obvious that mankind is destroying this heritage at an incredible speed. Not surprisingly, the biodiversity debate is laden with ethical, sometimes religious arguments. It was probably a biologist who has coined the saying "dont put price tags on my butterflies" ; however, economic reality puts "price tags" on biodiversity every day, mostly ignoring the moral considerations raised by experts.

The number of species endangered by human activities and the number of natural or semi-natural habitats being destroyed, fragmented or changed are constantly growing, thus destabilising ecosystems, causing the loss of vital resources together with genetic and cultural impoverishment.

Europe covers only 7% of the Earths land surface but contains a large biodiversity due to natural fragmentation by rivers, mountains, seas, the influences of glaciation, etc. The pressures on European biodiversity emanate from all sectors of society, with agriculture, forestry and transport being particularly responsible for habitat loss and fragmentation.

Measuring pressure on biodiversity, although an ambitious task, is essential to supply the controversial biodiversity debate with (hopefully) non-controversial, neutral and objective figures. Given the complexity of the issue, one should not expect perfect solutions. Describing threats to the "health" of ecosystems with just six indicators will resemble very much what a doctor would advise a human patient: "stop smoking, drink less, avoid fat meals and ride your bicycle every day." Most of the following indicators are of this rather general character. They are no substitute for a proper diagnosis, or a detailed plan to preserve a valuable habitat, but they may serve to publicly monitor the biggest threats to European biodiversity.

ACID RAINS

Other very dangerous pollutants are sulphur and nitrogen oxides. These gases are released by factories and power plants when fossil fuels are burned and by cars. These oxides reach high into the atmosphere and mix with water and other chemicals to form rain that can be as acid as vinegar. Acid rains are responsible for the decline of many forests. Tiny droplets of acid attack plant leaves, disrupting the production of chlorophyll. It also weakens the tree by altering the chemistry of the soil that surrounds its roots.

Acid falls down to earth as rain and snow. Black snow, as acid as vinegar, fell in Scotland in 1984. Acid rain affects everything it falls on. Rivers, lakes and forests are at risk throughout Europe and North America. In Sweden more than 18000 lakes have become acidic, 4000 of them very seriously indeed. This kills fish and drives out fish-eating wildlife.

Forests are particularly badly affected by acid rain and in many places previously green, luxuriant trees show bare branches at the top, stripped of foliage. In West Germany 50 per cent of trees are affected and, unless some curb is placed on pollution, the figure is certain to rise. In Austria, if nothing is done, scientists and environmentalists have predicted that there will be no trees left by the end of the century.

There is a possibility that damage to ecosystems from acid deposition may be more fundamental and long-lasting than was first believed. Scientists now report that acid rain leaches as much as 50 per cent of the calcium and magnesium from the forest soils. These minerals neutralise acids and are essential for plant growth. If soil chemistry is changed in this way, it may take many decades for all linked ecosystems to recover. Besides this, acid rain releases heavy metals and other toxic substances, providing a persistent source of toxicity to surrounding vegetation and aquatic life.

Buildings "die" too. Some of the most beautiful historic buildings in the world are being eaten away by the dilute acid, rained on them. Notre Dame, Cologne Cathedral and St Paul's Cathedral have all been damaged.

A major problem with air pollution is that it does not obey national boundaries. The planet's wind cycles and currents can carry pollution hundreds of miles away from its original source. So Britain is a large contributor to air pollution in Sweden and creates more for Norway than Norway does itself. The pollutants of the USA end up on the eastern coast of Canada.

Acid rain emerged as a concern in the 1960s with observations of dying lakes and forest damage in Northern Europe, the United States and Canada. It was one of the first environmental issues to demonstrate how the chief pollutants – oxides of sulphur and nitrogen – can be carried hundreds of miles by winds before being washed out of the atmosphere in rain, snow and fog.

As evidence grew of the links between air pollution and environmental damage, legislation to curb emissions was put in place. The 1979 Geneva Convention on Long-Range Transboundary Air Pollution set targets for reduction of sulphur and nitrogen emissions in Europe that have largely been achieved. The 1970 and 1990 Clean Air Acts have led to similar improvements in the USA.

Many nations have adopted air quality standards to safeguard the public against the most common pollutants. These include sulphur dioxide, carbon monoxide, suspended particulate matter, ground-level ozone, nitrogen dioxide and lead – all of which are tied directly or indirectly to the combustion of fossil fuels. Substantial investments in pollution control have lowered the levels of these pollutants in many cities of some developed countries. But poor air quality is still a major concern throughout the industrialised world.

Meanwhile, urban air pollution has worsened in most large cities in the developing world, a situation driven by population growth, industrialisation and increased vehicle use. Despite pollution control effects, air quality has approached the dangerous levels, recorded in London in the 1950s, in such megacities as Delhi, Jakarta and Mexico City.

In some parts of Asia, such as Southeast China, Northeast India, Thailand and the Republic of Korea, and in the Pacific region acid rain is now emerging as a major problem. In the Asia region the use of sulphur-containing coal and oil is very high. In 1990 34 million metric tons of sulphur dioxide were emitted there, which is over 40 per cent more, than in North America. The effects are already being felt in the agriculture. In India wheat growing near a power plant suffered a 49-per cent reduction in yield. Other ecosystems are also beginning to suffer. Pines and oaks in acid rain-affected areas of the Republic of Korea showed significant declines in growth rates since 1970. Many countries in the world are trying to solve the problem of air pollution in various ways, either by trying to burn fossil fuels more cleanly or by fitting catalytic converters to their cars, so fewer poisonous gases are produced. In some countries, like Sweden for example, new power plants use a method called fluidised bed combustion, which cuts sulphur emission down by 80 per cent. In Germany sulphurous smoke is sprayed with lime to produce gypsum, which is then used for building roads. Developing technologies like this may raise the price of electricity a little, but will save millions of trees, plants and animals and human health.

CARBON MONOXIDE

Carbon monoxide (CO) is a colourless, odourless and tasteless gas. Major sources include motor vehicles, home heating systems, and refuse burning. Outdoors, CO is generally a problem pollutant only in congested urban centres with high traffic density, especially in winter, as vehicle engines produce more pollution in cold weather.

In some traffic conditions, drivers and passengers in vehicles may be exposed to higher levels of CO than pedestrians in city streets. Malfunctioning or badly-vented heating furnaces, gas cooking stoves or woodstoves may also produce dangerous levels of CO in buildings. Other environments with an above-average risk of elevated CO levels include ice skating rinks, parking garages and traffic tunnels (WHO, 1999).

Background concentrations of CO in remote parts of the Northern hemisphere are around 0.1 ppm (WHO, 1999). Average annual concentrations in centres across Canada range from 0.5 to 1.0 ppm (Environment Canada, 2002). If inhaled, CO quickly enters the blood stream, where it reduces the ability of the blood to deliver oxygen to the organs and tissues of the body. The health impact of CO is most serious for those who suffer from cardiovascular disease, although healthy individuals are also affected at higher concentrations. Exposure to elevated levels is associated with numerous symptoms, including headache, fatigue, confusion, nausea, and impairment of vision. Work capacity, learning ability and the ability to perform complex tasks are reduced.

Nitrogen Dioxide

Nitrogen dioxide (NO Sulphur dioxide (SO_2) is a colourless gas with a characteristic smell, like that of a struck match. When levels reach between 0.3 to 1.0 parts per million (ppm), it is generally detectable as an acidic taste in the air. SO_2 dissolves very readily in water, eventually forming sulphuric acid. SO_2 is one of the most significant pollutants in many industrialised nations: it is often emitted as a result of burning sulphur-containing fuels, and by metal smelters and oil refineries.

Emissions of SO_2 from motor vehicles are relatively minor. Health effects associated with exposure to high concentrations of sulphur dioxide include breathing discomfort, respiratory illness, alterations in the lungs' normal defences, and aggravation of existing respiratory and cardiovascular disease. People with asthma, chronic lung or heart disease are the most sensitive to SO_2.

Apart from effects on human health, SO_2 can also directly damage the leaves of trees, and agricultural crops. SO_2 also has important indirect impacts by contributing to acid rain, which acidifies lakes and rivers, speeds up the corrosion of masonry and metal objects, and removes essential elements from the soil.

Sulphur dioxide also contributes to the formation of acid aerosols (fine acidic particles), which may have detrimental health effects. These small particles cause a visual haze, which affects the enjoyment of scenic vistas and has an effect on the earth's radiation balance.

Hydrogen Sulphide

Hydrogen sulphide (H_2S) is a colourless gas with a characteristic smell of rotten eggs. H_2S is flammable and can form an explosive mixture with air or oxygen. Common sources of H^2S include pulp and paper mills, oil refineries, steel processing plants, natural gas plants, and sewage treatment systems. Natural sources include sulphurous hot springs and volcanic vents, swamps and bogs.

Naturally-occurring sulphate-reducing bacteria can also produce H^2S in domestic water systems. The processes which produce hydrogen sulphide often generate other sulphur-containing compounds, such as methyl mercaptan and dimethylsulphide. These substances are also highly malodorous, and are often grouped together with hydrogen sulphide and referred to as total reduced sulphur compounds, or TRS. In the ambient air, the undesirable effects of H_2S are generally felt in terms of a bad odour. The odour threshold for H_2S is generally 1-5 ppb, but can be lower for some individuals. Background concentrations in unpolluted areas are around 0.1 ppb (WHO, 2003). There is some evidence that persistent exposure to unpleasant odours may have a significant effect on community health: review and monitoring of relevant health studies continues. There is also evidence that H_2S can contribute to an increased rate of spontaneous abortions when annual concentrations exceed approximately 3 ppb.

At higher concentrations (from 2 to approximately 100 ppm) hydrogen sulphide affects the mucous membranes, causing eye and throat irritation. Above this level H_2S becomes life-threatening as the sense of smell is impaired, removing the body's normal warning signal.

High concentrations of hydrogen sulphide can also cause paint discoloration and metal corrosion, and are damaging to plants. Life-threatening levels, in excess of 300 ppm, can exist in underground sewer pipe systems and occasionally above ground near sewer manholes, in situations where sewer gases are released from underground piping. During drilling for oil and gas, blow-outs can also occur with the release of large quantities of H_2S. Ambient concentrations in New Brunswick, even in the vicinity of large emission sources, seldom exceed 40 ppb (0.04 ppm).

Total Reduced Sulphur

Total Reduced Sulphur (TRS) is a family of odorous compounds including hydrogen sulphide, methyl sulphide, dimethyl sulphide and dimethyl disulphide. Some common sources of TRS include petroleum refineries, storage tanks for unrefined petroleum products, natural gas plants, petrochemical plants, sewage treatment facilities, and pulp and paper plants that use the kraft pulping process. Monitoring in the vicinity of kraft pulp mills in New Brunswick generally includes TRS. When evaluating monitoring data for TRS, reference may be made to standards or objectives for hydrogen sulphide. Although hydrogen sulphide may be a major component of TRS, the exact proportion is not known unless specific analysis for both H_2S and TRS is performed.

Ground-Level Ozone

Ozone (O_3) is a reactive, unstable form of oxygen. In very high concentrations, it is a bluish gas. The characteristic sharp smell of ozone may be recognised around electrical equipment such as motors or arc welders, where it is formed by the high temperature electric spark. In the concentrations found in outdoor air, even in the most severe pollution episodes, ozone is both colourless and odourless. Ozone is not emitted directly from smokestacks and exhaust pipes, but is formed in the air from other pollutants, most importantly nitrogen oxides (NO_2) and hydrocarbons or VOCs (such as solvent and gasoline vapours).

Vehicle exhaust contains all the necessary compounds to produce ozone, and, as a result, cars and trucks are important contributors to ground level ozone concentrations. Industrial emissions of NOx and hydrocarbons are also significant precursor emissions for ground level ozone.

Slow-moving air and strong sunshine greatly speed up the formation of ozone. As a result, ground level ozone is not typically a problem pollutant in the cooler months, but is likely to be more significant in hot, hazy summer weather. Such haze may build up over a period of days into a "photochemical smog".

However, the yellow or brownish colour of such smog is usually due to nitrogen dioxide and fine chemical particles, not the ground level ozone itself. Because the formation of ground level ozone depends on weather conditions, the severity of ozone pollution can be very variable from one year to the next.

In New Brunswick, significant ozone episodes occur on average about six times per summer, mainly affecting the southern part of the province. Much of this ozone originates from populated regions of the north-eastern United States, especially the Washington to Boston "corridor".

Ground-level ozone irritates the lungs, causing airway inflammation, and can make breathing difficult. Exposure to high concentrations can result in chest tightness, shortness of breath, reduced lung capacity and exercise performance, and coughing and wheezing. Recent health research suggests there is no clear "safe" threshold for ozone. Asthmatics appear to be more sensitive to repeat ozone exposures than healthy subjects. Concentrations as low as 60 ppb cause a measurable reduction in exercise performance.

There appears to be an additive or synergistic effect between ozone and some other principal air pollutants such as SO^2 and NO^2 (Health Canada, 1999). Ground level ozone can cause damage to agricultural crops which are ozone-sensitive, such as potatoes and tomatoes. It can also cause noticeable leaf damage in other species. Forest trees and other vegetation may be injured, or growth inhibited. Ozone also weakens rubber, and attacks metals and painted surfaces.

Ozone levels are often lower in city centres than in surrounding rural areas. This is because ozone reacts with nitrogen oxides, often found in relatively high concentrations in city cores due to emissions from motor vehicles. This effect is seen in Saint John, although to a lesser extent than in larger cities.

Total Suspended Particulate (TSP) Particles in the air (also called "particulates" by air pollution scientists) may either be of natural origin or the result of human activity. Some common natural sources and types of particulates include wind-blown soil dust, forest fires, sea salt, volcanoes, and plants (which produce pollen and spores). Human activities which generate particles include fuel combustion (domestic and industrial) and any other burning (e.g.waste incineration or slash burning); travel on dirt roads, construction work, and mining and quarrying. Dust on road and work sites is often wetted down to control particulate emissions. Other kinds of particles form in the air when gases such as sulphur dioxide (SO_2) and nitrogen oxides(NOx) react together. These particles are partly responsible for the yellowish "smog" sometimes seen over large cities.

"Total suspended particulate" (TSP) hasbeen a standard pollution measurement for many years, obtained by drawing a large volume of air through a special filter.This method makes no distinction between natural particles, such as pollen and spores, and particles which are emitted from vehicles or smokestacks.

People with existing breathing complaints such as asthma, bronchitis or emphysema, are liable to be adversely affected by high concentrations ofparticulates. Particulates can also cause corrosion and soiling of metalwork or other materials, damage vegetation, and reduce visibility.

PM 10 Particles smaller than 10 microns (a micron is a millionth of a metre, or a thousandth of a millimetre) are called PM 10, and are not filtered out of the air by the nose, throat and upper windpipe.

As a result, these very small particles can enter the lungs where they may irritate or otherwise have an adverse effect on the delicate membranes and air sacs which absorb oxygen. Other toxic effects may occur, depending on the chemical make up of the fine particles.

Because they can enter the lungs, PM 10 particles are sometimes referred to as 'inhalable' particles, and are measured separately from TSP. Many health studies have concluded that lung function, cardio-respiratory mortality, respiratory illness-related hospital admissions, and a variety of other health factors are related to ambient concentrations of PM10.

However, which particular aspect of the fine particles is causing these effects has not yet been fully resolved. Particle chemistry, and possibly acidity, may be important. As for ozone, there appears to be no clearly defined "safe" threshold for PM10, or PM2.5.

Recent studies also suggest that particles have effects on the immune system. PM 10 and PM 2.5 have been declared "toxic" under the Canadian Environmental Protection Act (Environment Canada, 2000). This means that these substances must undergo a detailed review and actions must be put in place to reduce their production and release into the environment.

In Canada, only the Greater Vancouver Regional District (GVRD) and the province of Newfoundland have a standard in place for PM 10. In the United States, standards exist both nationally and in California, which has the same standard as the GVRD. PM 2.5 Another standard specification for measuring particulates is PM 2.5, which refers to particles equal to or smaller than 2.5 microns in size.

These particles are thought to be of special significance in terms of health impacts, as they have a higher chance of entering and remaining in the lungs if inhaled, compared to PM 10 or larger particles. They are sometimes referred to as 'respirable particles'.

Particulate matter in the PM 2.5 size range is emitted from industrial and domestic combustion sources including slash burning, industrial processes (e.g. power plants, pulp mills, mining, smelting, and refining), open fires, woodstoves, forest fires, and internal combustion engines of all kinds. PM 2.5 also forms when other air pollutants react together in the air.

An important chemical reaction leading to the production of PM 2.5 is the combination of sulphur dioxide with ammonia, producing ammonium sulphate. Many other reactions are possible between natural atmospheric components and the air pollutants typically emitted in populated regions of North America and the world in general. Particles produced by this "gas-to particle conversion" are very small and contribute to haze over large areas. Transportation, industrial and other fuel combustion account for about 25% of the PM 2.5 emitted to the air.

The rest comes from so-called "open" sources, which include open burning, forest fires, road dust and construction. The estimated emissions from road dust and forest fires are very large, and dominate the inventory. Emissions from some source sectors are of greater concern than others in terms of health and other environmental effects. Recent studies indicate that particulate from combustion sources is more "biologically active" than road or other dust from terrestrial sources. Diesel exhaust is of special concern as it is thought to contribute significantly to the risk of lung cancer (USEPA, 1999).

Numerous studies worldwide continue to find a significant association between PM 2.5 concentrations and a wide range of health effects, including premature mortality, many cardio-pulmonary symptoms, reduced activity, increased physician visits and other forms of disability.

Monitoring technologies were introduced in the mid 1990s which allow the continuous measurement of PM 10 or PM 2.5. These units provide continuous, unattended recording and electronic logging and reporting functions.

Volatile Organic Compounds

VOCs are carbon-containing substances which are present in the air in gas form, usually at low concentrations (in the parts per billion range). Used in the context of air quality, VOCs are substances which often evaporate readily. Thousands of different VOCs exist in the air. However, many VOCs are of natural origin, and in terms of environmental protection, the VOCs which cause some kind of problem are most important. Everyday examples of VOCs which may be a concern in the environment include gasoline and organic solvents.

Other products such as paint, adhesives, and a variety of other household goods may contain solvents which evaporate and contribute to the VOC content of the air. VOCs are also emitted by a wide range of industrial sources, as well as by motor vehicles. Vehicles emit VOCs in their exhaust, and from fuel tanks. Modern vehicles have systems in place to minimize such emissions, but overall, transport-related emissions are still substantial. A large number of VOCs can be emitted by industrial processes (for example by the petrochemicals or plastics industries).

Some VOCs may be potentially toxic to humans, animals or vegetation (e.g. benzene, 1,3 butadiene, ethylene) or are of concern because they can react in the air to form other harmful substances. For example, ethylene and propylene (and many other VOCs) may take part in reactions which lead to the formation of ozone-containing smog. Some VOCs, such as various petroleum hydrocarbons, may contribute to undesirable odours. The VOC sampling program in New Brunswick concentrates on about 150 compounds which are of special interest, including known and suspected carcinogens such as benzene and 1,3 butadiene, and photochemically active VOCs which contribute to smog formation.

TYPES OF AIR POLLUTION MOST HAZARDOUS TO HUMANS

There are two forms of air pollution considered to be most harmful to humans by the American Lung Association. These are ozone, or smog, and particle pollution, or soot. Most ozone is formed by a chemical reaction between sunlight and the vapors emitted by the burning of carbon based or fossil fuels. Ozone pollution is generally highest during the sunniest months of the year, from May through October. This pollutant can cause short term health issues immediately following exposure, such as irritation to skin and the respiratory system, and long term exposure can lead to more serious health problems, such as impaired lung function, inflammation of the lung lining, and higher rates of pulmonary disease.

Particle Pollution

Particle pollution also takes a place at the top of the list of most dangerous to human health, and is very widespread throughout the environment. This type of air pollution consists of solid and liquid particles made up of ash, metals, soot, diesel exhaust, and chemicals. Particle pollution is produced by the burning of coal in power plants and other industries, and by the use of diesel fuel in passenger vehicles, cargo vehicles, and heavy equipment. Wood burning is a source of particle pollution, as are many of today's agricultural practices. Capable of triggering strokes, heart attacks, and irregular heart rates, particle pollution can be dangerous even in low concentrations. Lung cancer and premature birth have also been linked to exposure to particle pollution, and it is known to irritate respiratory conditions, including asthma, and cause coughing, wheezing, and even shorter life spans.

Other Common Air Pollutants

Other common air pollutants that can pose health risks to humans are carbon monoxide, nitrogen oxides, sulphur dioxide, and lead. Carbon monoxide is produced by the incomplete burning of fossil fuels in vehicles, home heating equipment, and industrial plants, among many other sources, and is a colorless and odorless gas, poisonous to humans and animals when inhaled. Nitrogen oxides are gases that contribute to smog and acid rain. Sulphur dioxide is produced by the burning of sulphur containing fuels like oil and coal, and can cause health issues, especially in those with existing heart or lung conditions. Lead is emitted into the air by vehicles and industrial sites, as well as by waste burning facilities. Lead is a neurotoxin when present in the body in high concentrations, and can cause immune issues, reproductive problems, kidney disease, and cardiovascular problems.

Greenhouse Gases

Perhaps the most publicized form of air pollution these days is the mixture of gases emitted into the air that are thought to be responsible for producing the greenhouse effect, leading to global warming and climate change. A certain percentage of greenhouse gases are produced by natural sources and are necessary to moderating the climate of the earth, making it possible for its life forms to survive. However, beginning with the Industrial Revolution, man has added to that production of greenhouse gases, primarily by the burning of fossil fuels. Among the most common of these are carbon dioxide, methane, and nitrous oxide.

Greenhouse gases collect in the atmosphere, forming a layer of reflective and absorbent materials that prevents some of the heat radiated by the sun from escaping the Earth's atmosphere, keeping the temperature sufficiently warm for plant and animal life to thrive. However, with the addition of man made greenhouse gases, too much heat can be reflected back into the atmosphere, giving rise to the current fears about global warming. Estimates made by the IPCC, or Intergovernmental Panel on Climate Change, predict that greenhouse gas emissions will double within the next 50 to 100 years at current rates of growth, leading to a variety of detrimental environmental effects. Among these are the melting of polar ice, raising ocean levels and flooding coastal and other low lying land areas. Increased storm activity and increased force and severity of hurricanes, cyclones, and tropical storms are among the possible consequences of climate change, as are severely altered ecosystems and extinctions of plant and animal species.

New Concern and New Hope for the Future

While the many types of air pollution that contaminate out air today are certainly of concern, awareness is growing about the danger they pose to our health and our planet. New regulations put into place over the past decade or two, such as the Clean Air Act and others, have significantly reduced the amounts of pollution pumped into the air we breathe every day. While there is much more to be done, environmentalists have managed to bring global warming and other environmental hazards to the forefront, gaining support from the public and the politically connected, advancing their cause in the halls of the United States government as well as in international forums.

MAJOR PROBLEMS OF AIR POLLUTION

Air pollution is one of the major problems of the present world. There are so many causes that can pollute the air completely and spread different types of dangerous diseases. On the basis of the different causes of air pollution it can be distinguished into different types. Mainly the type of the air pollution depends on some of the natural resources and most of them are composed of natural sources. Some important and common types that have great effect on the air of the environment and they also take part in spreading different types of dangerous diseases. Some types of air pollution are as follows:

SMOG

The first type of the air pollution is the smog. It is defined as when the smoke present in the atmosphere after emitting from different sources is combined with the fog present in the air, a mixture formed that is referred to as smog. Basically different types of factories or the industries are responsible for the formation of the smog. when the industries do their production from different materials, they can use different types of chemicals for the cleaning, refining or some kind of production processes, as a result these chemicals can produce different types of toxic materials that can emits in the form of the smoke from the chimney of the factory and form a bond of with the fog and cause different harmful diseases. Living in the smog is equal to the living with smokers; it can cause serious respiratory diseases.

Green House Effect

Another type of the air pollution is the green house effect. It is that type of air pollution that is formed due to the contamination of several important gases with the air. it is characterized when the gases called as green house gases when move upward and combine with the atmosphere and then return back to the earth and destroy different types of things such as crops, plants, human lives, livestock etc. These gases are basically six in number and they are; methane, sulphur, nitrogen, carbon monoxide, hydrogen and ozone. Basically the pollution is raised due to the burning of fossil fuel. It is very harmful for the human skin and can also cause some kind of cancer.

Accidental air Pollution

It is the type of pollution that is characterized due to the causes that are accidentally in nature. Commonly it is defined as the type of air pollution that is generated due to the different types of fuel consumption by the vehicles or when the forest are burnt different types of gases are evolved that are mixed with the air and pollute the air. Some times this pollution is also spread due to the plant leakage or different types of blasts in the furnaces of the manufacturing plants.

Industrial Air Pollution

Another type of air pollution that pollutes the environment as a result of the industrial processes is called as industrial pollution. Commonly it is characterized due to the working of the thermal plants and also the different plants that are used to manufacture different types of fertilizers or pesticides. The reactions that are used to produce different types of building material such as cement or steel etc. also encourage the production or toxic materials for producing air pollution. On the whole the air pollution due to the industrial wastes is called as industrial air pollution. Different type of atomic units also contributes in that type of pollution.

Transport Related Air Pollution

It is that type of air pollution that is characterized due to the smoke emitting by different types of vehicles used for transportation. As fuel such as petrol or diesel burnt in the engine can emit different types of poisonous gases in the form of smoke. This pollution can cause different types of harmful diseases.

CARBON DIOXIDE AND TEMPERATURE LEVELS

Since the late 1990s, perhaps the most vivid and compelling image of the connection between changes in CO2 and changes in global temperatures has been the chart—in a series of increasingly refined versions, now going back a million years—showing the two variables rising and falling together through a succession of ice ages. The rub has been that the changes in carbon dioxide have appeared to lag changes in temperature, rather than lead them (as one would expect if they were causing the temperature changes), and that the lags can be as long as thousands of years. In a paper that appeared on March 1 in *Science* magazine, a team of scientists report that using new techniques and reanalyzing data, they have virtually eliminated that puzzling temperature/CO_2 lag for the last ice age termination, the one most highly resolved..

The underlying problem has to do with uncertainties in estimation of annual changes in carbon dioxide levels. Yearly temperatures are inferred directly from changes in the isotopic composition of water deposited annually in snowfall; yearly accumulations are fairly easily distinguished because each year the top surface of the snow melts and then refreezes, forming a kind of crust called "firn." But the air bubbles in which carbon dioxide is trapped tend to diffuse through the crust, making it difficult to match up the bubbles wit the years in which they originally were trapped. As a companion commentary to the *Science* article explains, "Over the top 50 or 100 m of an ice sheet, the snowpack (firn) gradually becomes denser before it becomes solid ice containing air bubbles. Air diffuses rapidly through the firn, and the trapped air is therefore younger than the surrounding ice. In places with little snowfall, the age difference can be several thousand years. The age difference cannot be reconstructed perfectly, leading to uncertainty in the age of air..."

In the work reported on Friday, the multi-national team of European scientists used a proxy to better estimate the time of air bubble formation in the Antarctic core EPICA Dome C. Whereas the original analysis of that core had found changes in carbon dioxide lagging temperature changes by an average of 800 years in the last deglatiation, plus/minus 600 years, the new analysis halves the lag and cuts the uncertainty by a factor of three. "Their analysis indicates that CO2 concentrations and Antarctic temperature were tightly coupled throughout the deglaciation, within a quoted uncertainty of less than 200 years," says commentator Edward J. Brook, of Oregon State University, Corvallis.

How much of an impression will the new results make? Will they materially change the chemistry of the debate over human-induced climate change and climate policy? Doubtful.

For one thing, in part because of the complexity of the scientific methods used in both the original study and the new re-analysis, it will be easy for stubborn skeptics to believe that the scientists have simply picked a method that gives them the result they want. Second, much as one hates to trot our a tired cliche, the new results may raise more questions than they settle. Even if the changes in the two variables are indeed much more tightly linked, what co-factors are responsible for the whole pattern?

Brook puts it like this in the concluding paragraph of his commentary: "The ultimate question is what mechanisms influence both Antarctic climate and CO2 concentrations on such intimate timescales. Many have been discussed, and many are plausible, including changes in CO2 outgassing from the ocean due to changes in sea ice, changes in iron input to the ocean that influence CO2 uptake by phytoplankton, and large-scale ocean circulation changes that cause release of CO2 to the atmosphere. Deciding which are viable has proven difficult..."

THE EFFECTS OF AIR POLLUTION: ACID RAIN

Acid rain describes sulfuric and nitric acids deposited from the atmosphere. Often associated with precipitation, the term also applies to dry acidic materials. These acids commonly result from sulphur dioxide and nitrogen oxides reacting with moisture and other substances in the atmosphere. Although there are natural sources for these chemicals, much attention has been given to man-made sources, such as coal power plants. Acid rain is problematic due to acidification of soil, rivers, and lakes beyond the tolerance range of plants and animals. Acid rain can also erode man-made structures.

Acidifying Waters

Water resists rapid changes to pH — a measure of the acidity of a substance with lower numbers indicating stronger acidity. However, even this resistance is overcome by prolonged and persistent exposure to acid rain. Ecosystems within lakes and rivers may be vulnerable to acidification of the water in surprising ways. For example, mayflies die off at a pH of 5.5, while trout and perch can survive in much more acidic water. However, with the decline of mayflies and other insects, trout may have insufficient food to support their population. At pH 5 many fish eggs fail to hatch and juvenile fish tend to be more susceptible to acidity, impairing the fish population's continued health.

Forests

Direct contact with acid rain can weaken trees and destroy their leaves. This is especially true in high altitude forests where the trees are frequently immersed in an acid cloud. Acid rain can also harm trees in a more subtle fashion by reducing nutrient levels and increasing the level of toxic substances in the soil. The buffering capacity of soil varies greatly between different soil types, resulting in greater damage to forests in some areas than others, even though the acid rain exposure may be similar.

Cars

Many people take great pride in maintaining the appearance of their vehicle, but acid rain can literally erode the vehicle's protective coating. To counter these effects, automobile manufacturer's have begun coating new vehicles with acid-resistant paints.

Buildings

Limestone and marble construction materials are especially damaged by acid rain. This is due to the calcite mineral content in these materials that is easily dissolved away. This damage is readily seen in older stone buildings and monuments where carvings placed in the stone have eroded. Not all stone is susceptible. Granite and sandstone have a chemical composition that does not react with acid rain, though some types of sandstone contain carbonate, which will react.

Human Health

Physical contact with acid rain, either as droplets falling from the sky or from swimming in an acidic lake, has little direct impact on the health of humans. However, the pollutants responsible for the formation of acid rain are associated with an increase in respiratory disease and other illnesses. These pollutants may even infiltrate indoor spaces causing problems ranging from asthma to premature death. Laws such as the Clean Air Act strive to reduce the amount of pollution in the air. The U.S. Environmental Protection Agency reports that between 1980 and 2009, the national average for sulphur dioxide in the air decreased by 76 percent, and nitrogen dioxide decreased by 48 percent.

THE CIGARETTE IS A MAJOR SOURCE OF POLLUTION

The pollutants generated by the cigarette arise from the chemical process of burning organic matter, or*combustion* of tobacco and paper. Combustion processes, such as wood burning or waste incineration, emit thousands of pollutants, some of which are in the gas phase and some of which are in the form of small particles called *particulate matter*.

Particulate matter consists of millions of tiny particles of diverse chemical composition. Particulate matter from tobacco smoke includes many particles in the size range that reflects light, which explains why tobacco smoke is easily seen by the eye. In contrast to smoke particles, gases emitted by the cigarette such as benzene and carbon monoxide (CO) are invisible to the eye. Particles smaller than 2.5 micrometers ($PM_{2.5}$) are major components of cigarette smoke and can enter deep into the lung where they can cause serious health problems. To illustrate how small a PM $_{2.5}$ particle is, consider that 25,000 particles of this diameter, when placed side to side, can fit into 1 inch on a ruler.

Although a single cigarette is small in size and typically weighs less than 1 gram, a cigarette typically emits between 7 and 23 milligrams (mg) of $PM_{2.5}$ when it is smoked, depending on the manner of smoking and the brand. When people congregate in an airport baggage area or enter a smoking lounge where many brands are smoked, the average amount of $PM_{2.5}$mass emitted per cigarette is about 14 mg. Although 14 mg may not seem like a lot of mass emitted, each cigarette weighs only about 0.9 grams total, making it an extremely potent source of air pollution for its weight.

The particles emitted by each cigarette is really a large amount of particulate matter mass, causing extremely high indoor air pollutant concentrations when a cigarette is smoked at home or in a car. The chapter "Where does the smoke go?" presents calculations that you can do yourself to illustrate that a single cigarette smoked indoors is a potent source of exposure to toxic pollutants, causing concentrations indoors that are often higher than the federal air quality standards designed to protect public health in ambient air outdoors.

What Happens When a Person Smokes?

When a person smokes a cigarette, the part of the smoke that is inhaled directly into the lungs is called *mainstream* smoke. Pollutants inhaled in the mainstream smoke enter the lung directly and can be absorbed by the blood stream and body tissue. For example, inhaled carbon monoxide (CO) gas enters the blood stream where it ties to the human blood molecule (hemoglobin), thereby depriving the brain of oxygen as the blood enters the brain. Elevated CO in the blood may persist for many hours after the cigarette has been smoked, and it is possible to determine if a person has smoked a cigarette simply by measuring the elevated CO in a sample of the person's blood. Some portion of this inhaled smoke is exhaled by the smoker as processed mainstream smoke. More than half the pollutants emitted by a cigarette come not from the smoked end of the cigarette but from its other end – the cigarette's burning end – and are called *sidestream* smoke. Secondhand smoke, sometimes denoted as SHS, is a combination of both the exhaled mainstream smoke and the sidestream smoke emitted by the burning end of the cigarette, as well as any other smoke emitted from the end of the cigarette held by the smoker. Thus, secondhand smoke (SHS) is the total amount of pollution that leaves the immediate surroundings of the smoker.

Since each cigarette emits a large amount of fine particulate matter (7-23 mg) as SHS, and since this particulate matter comprises the visible part of the emitted smoke, some scientists believe the act of "smoking" should really be called "particling," a less romantic but scientifically accurate description of the activity of smoking. Tobacco company advertisements would be less exciting to young people and less likely to cause people to start smoking if these ads described the act of smoking as "particling."

Pollutants in Secondhand Smoke

Tobacco smoke air pollution is a mixture of more than 4,000 chemical by-products of tobacco combustion, 500 of which are in the gas phase. Of these byproducts of secondhand smoke, 172 are known toxic substances, many of which are regulated under existing clean air laws. In addition to 3 standard criteria air pollutants (carbon monoxide, particulate matter, and lead) and 33 hazardous air pollutants (HAPs) regulated in ambient air under the federal Clean Air Act, secondhand smoke contains 47 pollutants that are classified as hazardous wastes whose disposal in solid or liquid form is regulated by the Resource Conservation and Recovery Act, 67 pollutants known to be human or animal carcinogens, and 3 industrial chemicals regulated under the Occupational Health and Safety Act.

IMPACTS OF AIR POLLUTION & ACID RAIN ON VEGETATION

"Acid rain" is a general name for many phenomena including acid fog, acid sleet, and acid snow. Although we associate the acid threat with rainy days, acid deposition occurs all the time, even on sunny days.

Sulphur dioxide and nitrogen oxides both combine with water in the atmosphere to create acid rain. Acid rain acidifies the soils and waters where it falls, killing off plants. Many industrial processes produce large quantities of pollutants including sulphur dioxide and nitrous oxide. These are also produced by car engines and are emitted in the exhaust. When sulphur dioxide and nitrous oxide react with water vapour in the atmosphere, acids are produced. The result is what is termed acid rain, which causes serious damage to plants.

In addition, other gaseous pollutants, such as ozone, can also harm vegetation directly.

HOW ACID RAIN HARMS TREES

Acid rain does not usually kill trees directly. Instead, it is more likely to weaken the trees by damaging their leaves, limiting the nutrients available to them, or poisoning them with toxic substances slowly released from the soil. The main atmospheric pollutants that affect trees are nitrates and sulphates. Forest decline is often the first sign that trees are in trouble due to air pollution.

Scientists believe that acidic water dissolves the nutrients and helpful minerals in the soil and then washes them away before the trees and other plants can use them to grow. At the same time, the acid rain causes the release of toxic substances such as aluminium into the soil. These are very harmful to trees and plants, even if contact is limited. Toxic substances also wash away in the runoff that carries the substances into streams, rivers, and lakes. Fewer of these toxic substances are released when the rainfall is cleaner.

Even if the soil is well buffered, there can be damage from acid rain. Forests in high mountain regions receive additional acid from the acidic clouds and fog that often surround them. These clouds and fog are often more acidic than rainfall. When leaves are frequently bathed in this acid fog, their protective waxy coating can wear away. The loss of the coating damages the leaves and creates brown spots. Leaves turn the energy in sunlight into food for growth. This process is called photosynthesis. When leaves are damaged, they cannot produce enough food energy for the tree to remain healthy.

Once trees are weak, diseases or insects that ultimately kill them can more easily attack them. Weakened trees may also become injured more easily by cold weather.

How Air Pollution Harms Trees

Whilst acid rain is a major cause of damage to vegetation, air pollutants which can also be harmful directly. These include sulphur dioxide and ozone.

Sulphur Dioxide

Sulphur dioxide, one of the main components of acid rain, has direct effects on vegetation. Changes in the physical appearance of vegetation are an indication that the plants' metabolism is impaired by the concentration of sulphur dioxide. Harm caused by sulphur dioxide is first noticeable on the leaves of the plants. For some plants injury can occur within hours or days of being exposed to high levels of sulphur dioxide. It is the leaves in mid-growth that are the most vulnerable, while the older and younger leaves are more resistant. You can see the damage to coniferous needles by observing the extreme colour difference between the green base and the bright orange-red tips.

The effects of sulphur dioxide are influenced by other biological and environmental factors such as plant type, age, sunlight levels, temperature, humidity and the presence of other pollutants (ozone and nitrogen oxides). Thus, even though sulphur dioxide levels may be extremely high, the levels may not affect vegetation because of the surrounding environmental conditions. It is also possible that the plants and soils may temporarily store pollutants. By storing the pollutants they are preventing the pollutants from reacting with other substances in the plants or soil.

Ozone

The effects of ozone on plants have been investigated intensively for almost two decades. Studies made in controlled environment (CE) chambers, glasshouses and in the field, using open-topped chambers, have all contributed to the understanding of the mechanisms underlying ozone effects and their ultimate impact on vegetation. The biochemical mechanisms by which ozone interacts with plants have been intensively studied and, although the relative significance of different initial reactions remains unclear, there is a consensus that the key event in plant responses is oxidative damage to cell membranes. This primary oxidative damage results in the loss of membrane integrity and function, and in turn to inhibition of essential biochemical and physiological processes. A key target is photosynthesis, although ozone may also affect stomatal function and so modify plant responses to other factors, such as drought and elevated carbon dioxide. These changes result in reduced growth and yield in many plants. However, it is clear that such responses vary in magnitude between species and also between different cultivars within species. The mechanisms by which some species and genotypes are protected from ozone injury are not clear, but may include differences in uptake into the leaf or in the various components of antioxidant metabolism. Ozone may also increase the severity of many fungal diseases, while virus infections reduce the effects of ozone in some plants.

PAST AND PRESENT POLLUTION

Acid deposition and ozone exposure have increased considerably in the past 50 years in Asia, Europe and the US, with many reports of tree/forest decline and increased mortality. In general, the more highly polluted forests have the higher rate of decline and mortality. However, there has been no recent chronic deterioration in the UK of tree condition. Since the early 1990s, peak concentrations of ozone have been falling, whilst the large reduction in sulphur dioxide emissions since the 1970s has provided an opportunity for recovery of many plant species. By 2010, atmospheric sulphur dioxide concentrations in the UK should pose little or no threat to vegetation.

Acidification by Forestry

While forestry has long been considered to be adversely affected by air pollution and acid rain, recent studies show it to be part of the acidifying process. The rough canopies of mature evergreen forests are efficient scavengers of particulate and gaseous contaminants in polluted air. This results in a more acidic deposition under the forest canopies than in open land. Chemical processes at the roots of trees, evergreens in particular, further acidify the soil and soil water in forest catchments. When the forests are located on poorly buffered soils, these processes can lead to a significant acidification of the run-off water and consequent damage to associated streams and lakes.

ACID RAIN POSSESSES

Acid rain is a rain or any other form of precipitation that is unusually acidic, meaning that it possesses elevated levels of hydrogen ions (low pH). It can have harmful effects on plants, aquatic animals, and infrastructure. Acid rain is caused by emissions of sulphur dioxide and nitrogen oxide, which react with the water molecules in the atmosphere to produce acids. Governments have made efforts since the 1970s to reduce the release of sulphur dioxide into the atmosphere with positive results. Nitrogen oxides can also be produced naturally by lightning strikes and sulphur dioxide is produced by volcanic eruptions. The chemicals in acid rain can cause paint to peel, corrosion of steel structures such as bridges, and erosion of stone statues.

Definition

"Acid rain" is a popular term referring to the deposition of wet (rain, snow, sleet, fog, cloudwater, and dew) and dry (acidifying particles and gases) acidic components. Distilled water, once carbon dioxide is removed, has a neutral pH of 7. Liquids with a pH less than 7 are acidic, and those with a pH greater than 7 are alkaline. "Clean" or unpolluted rain has an acidic pH, but usually no lower than 5.7, because carbon dioxide and water in the air react together to form carbonic acid, a weak acid according to the following reaction:

H_2O (l) + CO_2 (g) $\rightleftharpoons$ H_2CO_3 (aq)

Carbonic acid then can ionize in water forming low concentrations of hydronium and carbonate ions:

H_2O (l) + H_2CO_3 (aq) HCO_3^-(aq) + H_3O^+ (aq)

However, unpolluted rain can also contain other chemicals which affect its pH. A common example is nitric acid produced by electric discharge in the atmosphere such as lightning. Acid deposition as an environmental issue would include additional acids to H_2CO_3.

History

The corrosive effect of polluted, acidic city air on limestone and marble was noted in the 17th century by John Evelyn, who remarked upon the poor condition of the Arundel marbles. Since the Industrial Revolution, emissions of sulphur dioxide and nitrogen oxides into the atmosphere have increased. In 1852, Robert Angus Smith was the first to show the relationship between acid rain and atmospheric pollution in Manchester, England.

Though acidic rain was discovered in 1853, it was not until the late 1960s that scientists began widely observing and studying the phenomenon. The term "acid rain" was coined in 1872 by Robert Angus Smith. Canadian Harold Harvey was among the first to research a "dead" lake. Public awareness of acid rain in the U.S increased in the 1970s after The New York Times published reports from the Hubbard Brook Experimental Forest in New Hampshire of the myriad deleterious environmental effects shown to result from it.

Occasional pH readings in rain and fog water of well below 2.4 have been reported in industrialized areas. Industrial acid rain is a substantial problem in China and Russia and areas downwind from them. These areas all burn sulphur-containing coal to generate heat and electricity.

The problem of acid rain has not only increased with population and industrial growth, but has become more widespread. The use of tall smokestacks to reduce local pollution has contributed to the spread of acid rain by releasing gases into regional atmospheric circulation. Often deposition occurs a considerable distance downwind of the emissions, with mountainous regions tending to receive the greatest deposition (simply because of their higher rainfall). An example of this effect is the low pH of rain which falls in Scandinavia.

History of acid rain in the United States

In 1980, the U.S. Congress passed an Acid Deposition Act. This Act established a 18-year assessment and research program under the direction of the National Acidic Precipitation Assessment Program (NAPAP). NAPAP looked at the entire problem from a scientific perspective. It enlarged a network of monitoring sites to determine how acidic the precipitation actually was, and to determine long term trends, and established a network for dry deposition. It looked at the effects of acid rain and funded research on the effects of acid precipitation on freshwater and terrestrial ecosystems, historical buildings, monuments, and building materials. It also funded extensive studies on atmospheric processes and potential control programs.

From the start, policy advocates from all sides attempted to influence NAPAP activities to support their particular policy advocacy efforts, or to disparage those of their opponents. For the U.S. Government's scientific enterprise, a significant impact of NAPAP were lessons learned in the assessment process and in environmental research management to a relatively large group of scientists, program managers and the public.

In 1991, DENR provided its first assessment of acid rain in the United States. It reported that 5% of New England Lakes were acidic, with sulfates being the most common problem. They noted that 2% of the lakes could no longer support Brook Trout, and 6% of the lakes were unsuitable for the survival of many species of minnow. Subsequent Reports to Congress have documented chemical changes in soil and freshwater ecosystems, nitrogen saturation, decreases in amounts of nutrients in soil, episodic acidification, regional haze, and damage to historical monuments.

Meanwhile, in 1989, the US Congress passed a series of amendments to the Clean Air Act. Title IV of these amendments established the Acid Rain Program, a cap and trade system designed to control emissions of sulphur dioxide and nitrogen oxides. Title IV called for a total reduction of about 10 million tons of SO_2 emissions from power plants. It was implemented in two phases. Phase I began in 1995, and limited sulphur dioxide emissions from 110 of the largest power plants to a combined total of 8.7 million tons of sulphur dioxide. One power plant in New England (Merrimack) was in Phase I. Four other plants (Newington, Mount Tom, Brayton Point, and Salem Harbour) were added under other provisions of the program. Phase II began in 2000, and affects most of the power plants in the country.

During the 1990s, research continued. On March 10, 2005, EPA issued the Clean Air Interstate Rule (CAIR). This rule provides states with a solution to the problem of power plant pollution that drifts from one state to another. CAIR will permanently cap emissions of SO_2 and NO_x in the eastern United States. When fully implemented, CAIR will reduce SO_2 emissions in 28 eastern states and the District of Columbia by over 70% and NO_x emissions by over 60% from 2003 levels.

Overall, the program's cap and trade program has been successful in achieving its goals. Since the 1990s, SO_2 emissions have dropped 40%, and according to the Pacific Research Institute, acid rain levels have dropped 65%

since 1976. Conventional regulation was utilized in the European Union, which saw a decrease of over 70% in SO_2 emissions during the same time period.

In 2007, total SO_2 emissions were 8.9 million tons, achieving the program's long term goal ahead of the 2010 statutory deadline.

The EPA estimates that by 2010, the overall costs of complying with the program for businesses and consumers will be $1 billion to $2 billion a year, only one fourth of what was originally predicted.

Human Health Effects

Acid rain does not directly affect human health. The acid in the rainwater is too dilute to have direct adverse effects. However, the particulates responsible for acid rain (sulphur dioxide and nitrogen oxides) do have an adverse effect. Increased amounts of fine particulate matter in the air do contribute to heart and lung problems including asthma and bronchitis.

Other Adverse Effects

Acid rain can also damage buildings and historic monuments and statues, especially those made of rocks, such as limestone andmarble, that contain large amounts of calcium carbonate. Acids in the rain react with the calcium compounds in the stones to create gypsum, which then flakes off.

$CaCO_3$ (s) + H_2SO_4 (aq) $CaSO_4$ (aq) + CO_2 (g) + H_2O (l)

The effects of this are commonly seen on old gravestones, where acid rain can cause the inscriptions to become completely illegible. Acid rain also increases the corrosion rate of metals, in particular iron, steel, copper and bronze.

Affected Areas

Places significantly impacted by acid rain around the globe include most of eastern Europe from Poland northward into Scandinavia, the eastern third of the United States, and southeastern Canada. Other affected areas include the southeastern coast of China and Taiwan.

GLOBAL AIR POLLUTION

As serious as water pollution is to the health and welfare of man, in many parts of the world air pollution represents an even more serious threat to human existence. Most of the major cities of the world and many rural areas as well now have serious air quality problems. The relative small (500,000 people) and modern capital of Malaysia, Kuala Lumpur, a beautiful city surrounded by luxuriant green hills, now has an emerging air pollution problem as its economy begins to prosper and expand. It arises from rapidly increasing automobile, truck and bus traffic, and burgeoning industrial development.

Even the remote capital of Nepal, Kathmandu, which is nestled in the Himalayan mountains north of India, began to show symptoms of air pollution in 1971. Famous for its clear mountain air and startling views of the high Himalaya, the valley of Kathmandu is often hazy with exhaust smoke from rapidly increasing automobile, truck, bus, and airplane traffic. The valley and city of Kathmandu are vulnerable to such pollution because of a natural air inversion a layer of warm tropical air over cold mountain air resting in an enclosed and densely populated valley. Unless control measures are taken now, the magnificent scenery which brings tourists to Nepal, will be viewed through a polluted haze in future years.

Ten years ago many engineers and environmental scientists insisted that there were few if any demonstrable ill effects on human health from the levels of air pollution then existing in most cities. They admitted exceptions to this, of course, but these were primarily confined to critical situations such as the famous London smog of 1952 in which 4,000 to 5,000 people died from respiratory distress in a persistent smog, or the crisis in Donora, Pennsylvania in 1948, in which hundreds of people suffered similar fatalities during and after a severe smog condition which hung over the city for several weeks. Other than these rare circumstances, there was little evidence that chronic levels of urban air pollution represented a public health problem.

Now there is abundant evidence that the levels of air pollution in many cities do represent a major medical problem. This evidence comes from both experimental research with animals and clinical studies in human health. In fact, the health hazards of air pollution in the United States, Japan and parts of Europe are now more clearly documented than are those of water pollution. This is not true, of course, throughout much of Latin America, Africa, and Asia, where impure water still represents a major health problem.

Some of the first clues to the health hazards of air pollution came from clinical observations of increasing rates of emphysema, chronic bronchitis, and respiratory distress in city dwellers, and from experimental studies on laboratory animals exposed to air pollutants. In the latter category of experimental studies, it was noted in Los Angeles several years ago that laboratory mice exposed to ambient air pollution (*i.e.*, normally existing air pollution) developed significant pathology compared to control mice in clean air. Aging inbred mice showed increased frequency of pulmonary adenoma, and one strain of rice showed increased mortality of young adult males. Severe smog episodes in Los Angeles caused basic changes in the cellular structure of lung tissue in these animals. Guinea pigs and rabbits developed altered hormone excretion patterns and differential enzyme levels in blood serum in contrast to clean air controls. Pathologic effects of air pollution have also been clearly demonstrated in people. In one study in New York City, children under 8 years of age showed a prevalence of respiratory symptoms directly related to levels of particulate matter and carbon monoxide. In many cities along the Eastern seabord, increasing evidence of dyspnea, bronchitis, cough, sputum production, wheezes, eye irritations and general malaise was elicited as air pollution levels increased. In Los Angeles, a correlation was shown between carbon monoxide pollution levels and case fatality rates in patients with heart trouble. Sulphur dioxide is one of the common gaseous air pollutants which is most injurious to human health. It irritates respiratory epithelium and impairs normal beathing. Most cities have SO_2 levels less than 0.5 ppm, and human effects are not prominent until 0.8 or 1.0 ppm are attained. Frequently, ambient levels in cities exceed 0.8 ppm when stagnant air remains for several days and gaseous pollutants accumulate. Investigators have shown correlations between photochemical air pollutants and respiratory distress, emphysema and susceptibility to respiratory infection. Chronic as well as acute effects have been documented. In the industrial complex of Bayonne and Elizabeth, New Jersey, the death rate from respiratory cancer in males was 35 per cent higher in an area of high air pollution, compared to a similar population living in a lower air pollution environment only a few miles away.

On Staten Island, the rate of lung cancer in women in polluted areas was shown to be twice that of women in clear areas. Other studies have shown that urban air pollution increases the rate of lung cancer in men three times above that of rural men. This type of evidence could be cited in considerable detail to leave little doubt that air pollution is detrimental to human health. In fact, respiratory ailments related to air pollution—emphysema, chronic bronchitis, lung ancer, and severe asthma—are among the most rapidly increasing health problems in industrialized nations. Various disease problems associated with air pollution were reviewed by Lave and Seskin who also evaluated the economic costs of respiratory disease attributable the air pollution. They estimated that a 50 per cent reduction in US urban air pollution would save the nation $2,080 million dollars per year in health and medical costs.

Air pollution is as complex in origin and type as water pollution. It has been estimated that 164 million metric tons of pollutants enter the United States air every year. This pollutant load is composed of a wide range of particulate matter, suspended particles, and gases. In New York City alone, the daily emission of air pollutants includes 3,200 tons of SO_2, 4,200 tons of CO, and 280 tons of particulate dirt. In Philadelphia in 1959, daily releases of air pollutants amounted to 830 tons of SO_2, 300 tons of NO_2, 1,350 tons of hydrocarbons, and 470 tons of particulates.

Particulate matter fall-out affords a dramatic example of dirty air. In many of the world's cities, the average daily air pollution fall-out is in the order 0.5 to 3.0 tons per square mile per day. Some cities have over 4 tons per square mile per day. In Pittsburgh, fall-out was reduced from over 5.5 tons per square mile per day to less than 0.9 tons by a vigorous clean air and smoke abatement programme. Up to one per cent of urban dust may be lead, which is toxic to humans in fairly low concentrations.

Yet particulate fall-out, while it may be one of the most dramatic forms of air pollution, is certainly not necessarily the most harmful. Most particles which fall out are over 100 micra in diameter, and these seldom if ever reach the alveolar tissue of the lung where irritation occurs. Fine suspended particles and gaseous substances are the primary agents of respiratory distress.

One of the most persistent forms of air pollution which has not responded well to control measures is automobile exhaust. In the late 1960's, over 90 million motor vehicles in the United States produced 66 million tons of carbon monoxide, and 20 million tons of other air pollutants per year. Although various devices have been developed to reduce exhaust emission, these devices are not consistently well maintained by the public.

Los Angeles, which has mounted an outstanding campaign of industrial air pollution control, still faces a major smog problem from automobile exhaust. The city has 4,000,000 automobiles for 3,000,000 people. Two-thirds of the area of the central city has been taken over by the automobile in the form of parking space, streets and freeways. One study in Los Angeles showed a direct correlation between air pollution and the frequency of motor vehicle accidents.

In the air on crowded freeways, carbon monoxide levels may reach 400 parts per million. Automobile drivers thought to be responsible for accidents showed elevated carbon monoxide levels in their blood. Many experts predict that the internal combustion engine must be either outlawed or significantly altered within the next few years if we are to avert air pollution tragedies in our cities.

There is no doubt that air pollution is a very significant and increasing factor in environmental deterioration. Airplane pilots who have flown for twenty or thirty years report a great increase in "ground haze" and air pollution domes encapsulating cities. Whereas cities were often seen from aerial distances of 30 to 40 miles some years ago, they are now usually enshrouded by air pollution and not visible from more than 5 or 10 miles.

In many areas, air pollution has caused dramatic injury to plants, both agricultural crops and natural plant communities. Citrus groves, truck garden crops of lettuce, tomatoes, onions and celery, field crops of alfalfa, sweet corn, and tobacco, and even forests of pine, spruce, and deciduous trees have all fallen victims to air pollution in various parts of the world.

Air pollution also takes its toll of buildings and other man-made objects. When moisture accumulates in polluted air, the oxides of sulphur, carbon and nitrogen form weak sulfuric acid, carbonic acid and nitrous acid which are corrosive to metals, stone, paint, rubber, textiles and even some plastics. One study estimated that air pollution in 1960 cost the average American family $800 per year in property damage. Current projections of this figure would certainly put these costs well over $1000 per family per year.

Throughout Europe, many famous buildings, monuments and art treasures of former centuries are deteriorating at an alarming rate due to the erosional effects of air pollution. In Athens, the President of the Greek Academy of Sciences, estimated in 1971 that the Parthenon on top of the Acropolis has deteriorated more in the last 50 years than in the previous 2000 years. Athens is normally blessed with clean fresh air, but this depends entirely on sea breezes. On still, windless days, a noxious air pollution haze quickly forms, even shrouding out the Aegean Sea near Pathens.

A serious possibility of world-wide air pollution that was first detected in the late 1960s is the occurrence of stratospheric pollution and a global air pollution veil; that is, a ring of air pollution circling the globe in the northern hemisphere around the latitudes of the US, Europe and Japan. Such a global veil has been detected by satellite photos, and confirmed by both Russian and US scientists. The formations have been temporary, and apparently occurred during periods of unusually stable currents of air circulation around the world. In other words, pollutants were being carried intercontinenally, so that air entering North America contained Japanese contaminants, and air entering Europe contained North American pollutants. Normally, the oceanic mixing of air would break up pollution bands. This is not a major problem now, but it is a serious portent of things to come. Certainly, many types of air pollution can be controlled by modern technology, but the costs must ultimately be borne by the public through higher prices for industrial goods, higher taxes, reduced profit margins in industry, and more careful monitoring of automobile exhaust control systems. In many cases of environmental quality control, the ultimate responsibility rests with the citizenry at large through established routes of political process.

The present cost of air pollution in terms of ill health, agricultural damage, and accelerated deterioration of construction materials and personal goods is very great, but it is so diffuse and indirect that we are not aware of the great price we already are paying for it. When we reach the point where the cost of air pollution is greater than the cost of controlling it, the public must demand appropriate governmental action and be willing to support it.

Bibliography

Alka Pareek: *Environment and Nutritional Disorders*, Aavishkar Pub, Delhi. 2003.

Anfinsen CB: *Advances in Protein Chemistry*. New York: Academic Press. 1972.

Banwari Lal: *Environmental Microbiology*, Cyber Tech Pub, Delhi, 2009.

Burch, William R. Jr.: *Human Ecology: An Environmental Approach*, Belmont, Duxbury Press, 1994.

Cathy Solheim: *Environmental Justice: Issues, Policies, and Solutions*, Washington, D.C., Island Press, 1995.

Chanderlekha Goswami: *Biochemistry and Instrumentation*, Manglam Pub, Delhi, 2011.

Costanza, R.: *Ecological Economics*, Columbia Uni. Press., New York, 1991.

Daubenmire, F.: *Plants and Environment*, New York, Wiley, 1947.

Dereniak, E. and D. Crowe, *Optical Radiation Detectord*, John Wiley and Sons, 1984.

Florkin, M.: *Comprehensive Biochemistry: Amsterdam*, Elsevier, 1975.

Gay, Kathlyn: *Saving the Environment: Debating the Costs*, New York, Franklin Watts, 1996.

Gedicks, Al.: *The New Resource Wars: Native and Environmental Struggles Against Multinational Corporations*, Boston, South End Press, 1993.

Goel, A K : *Basic Concept of Animal Chemistry*, Pearl Books, Delhi, 2008.

Hill, David: *The Quality of Life in America; Pollution, Poverty, Power, and Fear*, New York, Holt, Rinehart and Winston, 1973.

Jack Simons: *An Introduction to Theoretical Chemistry*, Cambridge Univ Press, Delhi, 1998.

Jasra, O.P. : *Environmental Biochemistry*, Sarup & Sons, Delhi, 2002.

Kruvant, W.J.: *People, Energy, and Pollution*, Cambridge, Ballinger Publishing, 1975.

LaBalme, J.: *A Road to Walk, A Struggle for Environmental Justice*, Durham, Regulator Press, 1987.

Lodish, H.: *Molecular Cell Biology*, New York, Freeman, 2000.

Logan John R. and Harvey L. Molotoch: *Urban Fortunes: The Political Economy of Place*, Berkeley, University of California Press, 1987.

Mahboob, Syed : *Handbook of Fruit and Vegetable Products*, Agrobios, Delhi, 2008.

Mandelker, D.R.: *Environment and Equity: A Regulatory Challenge*, New York, McGraw Hill, 1981.

Meyer, Art: *Earth Keepers: Environmental Perspectives on Hunger, Poverty, and Injustice*, Scottsdale, Herald Press, 1991.

Middleton, Nick: *Atlas of Environmental Issues*, New York, Facts on File, 1989.

Nelson DL, Cox MM: *Lehninger's Principles of Biochemistry*, New York, New York: W. H. Freeman and Company, 2005.

Pathak, N : *Dictionary of Food and Nutrition*, Mohit Pub, Delhi, 2005.

Patricia Trueman: *Nutritional Biochemistry*, MJP Pub, Delhi, 2007.

Paul Mohai: *Race and the Incidence of Environmental Hazards: A Time for Discourse*, Boulder, Westview Press, 1992.

Prasad, S.K. : *Biochemistry of Proteins*, Discovery, Delhi, 2010.

Raff, K. Roberts, and J. Watson, *Molecular Biology of the Cell*, New York: Garland Publishing, 1989.

Ram Naresh Mahaling: *Basics of Biochemistry*, Anmol, Delhi, 2008.

Ruchi Gautam: *A Textbook of Microbiology*, Arise, Delhi, 2007.

Sachs, Aaron: *Eco-Justice: Linking Human Rights and the Environment*, Washington, World Watch Institute, 1995.

Sarath Chandra Bose: *Biochemistry : A Practical Manual*, Pharma Med Press, Delhi, 2010.

Sayler, G.S., Fox, R., and Blackburn, J.W.: *Environmental Biotechnology for Waste Treatment*, Plenum Press, New York, 1991.

Shubhrata R. Mishra: *Plant Biochemistry*, Discovery, Delhi, 2010.

Singh, Mahinder: *A Textbook of Biochemistry*, Dominant Pub, Delhi, 2011.

Sontheimer, Sally: *Women and the Environment: Crisis and Development in the Third World*, New York, Monthly Review Press, 1991.

Suri, Nitin : *Molecular Biology and Biochemistry*, Oxford Book Company, Delhi, 2010.

Switzer, R.L.: *Experimental Biochemistry*, New York, W.H. Freeman and Company, 1977.

Szasz, Andrew: *Ecopopulism: Toxic Waste and the Movement for Environmental Justice*, Minneapolis, University of Minnesota Press, 1994.

Ullrich M *Bacterial Polysaccharides: Current Innovations and Future Trends* Caister Academic Press, 2009.

usubel, F.M.: *Current Protocols in Molecular Biology*, New York, John Wiley and Sons, 1989.

Wade, J. L.: *Society and Environment: The Coming Collision*, Boston, Allyn and Bacon, 1972.

Wenz, Peter S.: *Environmental Justice*, Albany, State University of New York Press, 1988.

Wood, E.J. Ferguson: *Marine Microbial Ecology*, London, Chapman and Hall, 1965.

Zupan, J.M.: *The Distribution of Air Quality in the New York Region*, Baltimore, Johns Hopkins University Press, 1973.

Index

❑❑❑